GPS for Land Surveyors

Second Edition

Jan Van Sickle

CRC PRESS

Boca Raton   London   New York   Washington, D.C.

Published in 2001 by
Taylor & Francis Group
6000 Broken Sound Parkway NW, Suite 300
Boca Raton, FL 33487-2742

Published in Great Britain by
Taylor & Francis Group
2 Park Square
Milton Park, Abingdon
Oxon OX14 4RN

Library of Congress Cataloging-in-Publication Data

Van Sickle, Jan
   GPS for land surveyors / by Jan Van Sickle.—2nd ed.
     p. cm.
   Includes bibliographical references and index.
   ISBN 1-57504-075-1
   1. Artificial satellites in surveying. 2. Global positioning system. I. Title
TA595.5 .V36 2001
526.9'82—dc21   2001022387

Taylor & Francis Group
is the Academic Division of T&F Informa plc.

Visit the Taylor & Francis Web site at
http://www.taylorandfrancis.com

GPS   vors

To

Heidi Van Sickle

About the Author

Jan Van Sickle, P.L.S. has 35 years experience as a land surveyor. He was privileged to supervise surveys with the first commercially available Global Positioning System receivers in the early 1980s and has worked with GPS ever since. He is the author of *GPS for Land Surveyors* and *1001 Solved Surveying Fundamentals Problems*. The latter is quoted in the feature Surveying Solutions each month in *POB* magazine. He has taught the Advanced Surveying Program at the Denver Institute of Technology and GPS seminars throughout the United States, and is a licensed surveyor in California, Colorado, and Oregon. He is currently the GIS Department Manager at Qwest Communications.

Preface

Five years ago, when this book was first published, I was not prepared for the wonderful response. When I was asked to teach Global Positioning System (GPS) seminars based on the book around the country and meet some of the surveyors who use it, I knew I would learn a lot. I certainly did, and this revision is built on what I learned from them and from the many phone calls, letters, and E-mails I have received asking questions and making suggestions. It's been great, and more than I could have hoped for. I want to thank everyone who has contributed to this new, improved version.

This book is intended to be useful to surveyors who have ventured into GPS surveying. I hope it is also helpful to engineers and others for whom GPS may not be a primary tool, but who find themselves in need of some additional understanding of the technology.

The breadth of knowledge required for modern surveying is increasing all the time and that is especially true in GPS. But surveyors have been there from the very beginning of the Global Positioning System, and that is now more than 20 years ago. The system developed by the Department of Defense with military applications in mind is attracting users from every imaginable discipline, but it is still surveyors out there pushing its limits.

Like most computerized equipment, the prices of GPS receivers are going down, just as their capabilities improve. With the full constellation of satellites in place, the convenience of using GPS has never been better. It has truly become a 24-hour worldwide system now. And it can be applied to a wide range of real-life everyday fieldwork with good results. The potential of GPS has always been great, but potential is now reality.

Still, getting full advantage from GPS takes some doing. For example, it isn't hard to operate a GPS receiver—matter of fact, most of them are so user-friendly you don't need to know the first thing about GPS to make them work; that is, until they don't. Getting coordinates from a GPS receiver is usually a matter of pushing a few buttons, but knowing what those coordinates are, and more importantly, what they aren't, is more difficult.

Many surveyors feel ill prepared for the use of GPS in their everyday work; the subject seems too complex, the time to digest it too long, and the books and seminars on the subject are either too complicated or oversimplified.

This book has been written to find a middle ground. It is an introduction to the concepts needed to understand and use GPS, not a presentation of the latest research in the area. An effort has been made to explain the progression of the ideas at the foundation of GPS, and get into some of the particulars too.

Finally, this is a practical book, a guide to some of the techniques used in the performance of a GPS survey. From its design through observation, processing, and RTK (real-time kinematic), some of the aspects of a GPS survey are familiar to surveyors, some are not. This book is about making them all familiar.

Contents

1

The GPS Signal

GLOBAL POSITIONING SYSTEM (GPS) SIGNAL STRUCTURE

GPS and Trilateration

GPS can be compared to trilateration. Both techniques rely exclusively on the measurement of distances to fix positions. One of the differences between them, however, is that the distances, called ranges in GPS, are not measured to control points on the surface of the earth. Instead they are measured to satellites orbiting in nearly circular orbits at a nominal altitude of about 20,183 km above the earth.

A Passive System

The ranges are measured with signals that are broadcast from the GPS satellites to the GPS receivers in the microwave part of the electromagnetic spectrum; this is sometimes called a passive system. GPS is passive in the sense that only the satellites transmit signals; the users simply receive them. As a result, there is no limit to the number of GPS receivers that may simultaneously monitor the GPS signals. Just as millions of television sets may be tuned to the same channel without disrupting the broadcast, millions of GPS receivers may monitor the satellite's signals without danger of overburdening the system. This is a distinct advantage, but as a result, GPS signals must carry a great deal of information. A GPS receiver must be able to gather all the information it needs to determine its own position from the signals it collects from the satellites.

Time

Time measurement is essential to GPS surveying in several ways. For example, the determination of ranges, like distance measurement in a modern trilateration survey, is done electronically. In both cases, distance is a function of the speed of light, an electromagnetic signal of stable frequency and elapsed time. In a trilateration survey, frequencies generated within an electronic distance measuring device, EDM, can be used to determine the elapsed travel time of its signal because the signal bounces off a reflector and returns to where it started. But the signals from a GPS satellite do not return to the satellite, they travel one way, to the receiver. A clock in the satellite can mark the moment the signal departs and a clock in the receiver can mark the moment it arrives. And since the measurement of the ranges in GPS depends on the measurement of the time it takes a GPS signal to make the trip from the satellite to the receiver, these two widely separated clocks must communicate with each other. Therefore, the GPS signal itself must somehow tell the receiver the exact time it left the satellite.

Control

Both GPS surveys and trilateration surveys begin from control points. In GPS the control points are the satellites themselves; therefore, knowledge of the satellite's position is critical. Measurement of a distance to a control point without knowledge of that control point's position would be useless. It is not enough that the GPS signals provide a receiver with information to measure the range between itself and the satellite. That same signal must also communicate the position of the satellite, at that very instant. The situation is complicated somewhat by the fact that the satellite is always moving.

In a GPS survey, as in a trilateration survey, the signals must travel through the atmosphere. In a trilateration survey, compensation for the atmospheric effects on the EDM signal, estimated from local observations, can be applied at the signal's source. This is not possible in GPS. The GPS signals begin in the virtual vacuum of space, but then, after hitting the earth's atmosphere, they travel through much more of the atmosphere than most EDM signals. Therefore, the GPS signals must give the receiver some information about needed atmospheric corrections.

It takes more than one measured distance to determine a new position in a trilateration survey or in a GPS survey. Each of the several distances used to define one new point must be measured to a different control station. For trilateration, three distances are adequate for each new point. For a GPS survey the minimum requirement is a measured range to each of at least four GPS satellites.

Just as it is vital that every one of the three distances in a trilateration is correctly paired with the correct control station, the GPS receiver must be able to match each of the signals it tracks with the satellite of its origin. Therefore, the GPS signals themselves must also carry a kind of satellite identification. To be on the safe side, the signal should also tell the receiver where to find all of the other satellites as well.

To sum up, a GPS signal must somehow communicate to its receiver: (1) what time it is on the satellite, (2) the instantaneous position of a moving satellite, (3) some information about necessary atmospheric corrections, and (4) some sort of satellite identification system to tell the receiver where it came from and where the receiver may find the other satellites.

The Navigation Code

How does a GPS satellite communicate all that information to a receiver? It uses codes. Codes are carried to GPS receivers by two carrier waves. A carrier wave has at least one characteristic such as phase, amplitude, or frequency that may be changed—modulated—to carry information. For example, the information, the music or speech, received from an AM radio station is placed on the carrier wave by amplitude modulation, and the information on the signal from an FM radio station is there because of frequency modulation. The two GPS carriers are radio waves too. They come from a part of the L-band. The L-band is a designation that includes the ultrahigh radio frequencies from approximately 390 MHz to 1550 MHz. Actually, the so-called L1 GPS frequency is a bit higher than the strict L-band definition, as you will see.

Wavelength

A wavelength with a duration of 1 second, known as 1 cycle per second, is said to have a frequency of 1 hertz (Hz) in the International System of Units (SI). A frequency of 1 Hz is rather low. The lowest sound human ears can detect has a frequency of about 25 Hz. The highest is about 15,000 hertz, or 15 kilohertz (kHz).

Most of the modulated carriers used in EDMs and all those in GPS instruments have frequencies that are measured in units of a million cycles per second, or megahertz (MHz). The two fundamental frequencies assigned to GPS are called L1 at 1575.42 MHz and L2 at 1227.60 MHz.

Codes

GPS codes are binary, strings of zeroes and ones, the language of computers. The three basic codes in GPS are the precise code, or P code; the coarse/acquisition code, or C/A code; and the Navigation code. There are a few related secondary codes, which will be discussed later.

The Navigation code has a low frequency, 50 Hz, and is modulated onto both the L1 and L2 carriers. It communicates a stream of data called the GPS message, or Navigation message (Figure 1.1). This message is 1500 bits long, divided into five subframes with 10 words of 30 bits each. These subframes are the vehicles for telling the GPS receivers some of the most important things they need to know.

The accuracy of some aspects of the information included in the Navigation message deteriorates with time. Fortunately, mechanisms are in place to prevent the message from getting too old. The message is renewed each day by government uploading facilities around the world, or by the satellites themselves, in the case of the Block IIR satellites, the R stands for replenishment—more about that later. These installations, along with their tracking and computing counterparts, are known collectively as the *Control Segment* (see Chapter 3).

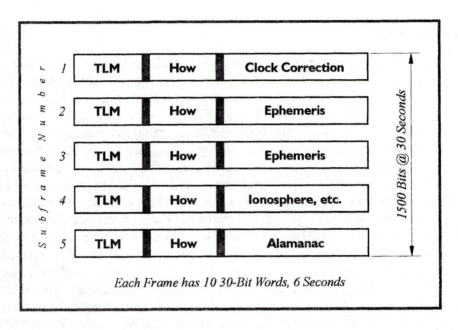

Figure 1.1. Navigation Message.

GPS Time

One example of time-sensitive information is found in subframe 1 of the Navigation message. Using a standard time scale called *GPS Time*, the message contains information needed by the receiver to correlate its clock with that of the clock of the satellite. But the constantly changing time relationships in GPS can only be partially defined in this subframe. It takes more than a portion of the Navigation message to define those relationships. In fact, the rate of the GPS Time standard is defined outside the system.

For example, the rate of GPS Time is kept within 1 microsecond of the rate of the worldwide time scale, which is called Coordinated Universal Time, and is abbreviated UTC. The rate of UTC, determined by more than 150 atomic clocks located around the globe, is more stable than the rotation of the earth itself. This causes a discrepancy between UTC and the earth's actual motion. This difference is kept within 0.9 seconds by the periodic introduction of leap seconds in UTC. But since GPS is not earthbound, leap seconds are not used in GPS Time. GPS Time was identical to UTC on midnight January 5, 1990. Since then, many leap seconds have been added to UTC but none have been added to GPS Time. This complicates the relationship between UTC and GPS Time. Even though their rates are virtually identical, the numbers expressing a particular instant in GPS Time are different by some seconds from the numbers expressing the same instant in UTC.

Satellite Clocks

Each GPS satellite carries its own onboard clocks in the form of very stable and accurate atomic clocks regulated by the vibration frequencies of the atoms of two elements. Two of the onboard clocks are regulated by cesium and two are regulated by rubidium, except on the Block IIR satellites, where all the clocks are rubidium. Since the clocks in any one satellite are completely independent from those in any other, they are allowed to drift up to one millisecond from the strictly controlled GPS Time standard. Instead of constantly tweaking the satellite's onboard clocks to keep them all in lockstep with each other and with GPS Time, their individual drifts are carefully monitored by the government tracking stations of the Control Segment. These stations record each satellite clock's deviation from GPS Time, and the drift is eventually uploaded into subframe 1 of each satellite's Navigation message, where it is known as *the broadcast clock correction.*

A GPS receiver may relate the satellite's clock to GPS Time with the correction given in the broadcast clock correction of its Navigation message. This is obviously only part of the solution to the problem of directly relating the receiver's own clock to the satellite's clock. The receiver will need to rely on other aspects of the GPS signal for a complete time correlation.

The drift of each satellite's clock is not constant. Nor can the broadcast clock correction be updated frequently enough to completely define the drift. Therefore, one of the 10 words included in subframe 1 provides a definition of the reliability of the broadcast clock correction. This is called *AODC,* or Age of Data Clock.

Another example of time-sensitive information is found in subframes 2 and 3 of the Navigation message. They contain information about the position of the satellite, with respect to time. This is called the satellite's *ephemeris.*

The Broadcast Ephemeris

The satellite's ephemeris is given in a right ascension, RA, system of coordinates. There are six orbital elements. They are: the semimajor axis of the orbit, the eccentricity of the orbit, the right ascension of its ascending node, the inclination of its plane, the argument of its perigee, and the true anomaly.

This information is contained in these two subframes and contains all the information the user's computer needs to calculate earth-centered, earth-fixed, WGS 84 coordinates of the satellite at any moment. The broadcast ephemeris, however, is far from perfect.

For example, the broadcast ephemeris is expressed in parameters that appear Keplerian. (Its elements are named for the seventeenth century German astronomer Johann Kepler.) But in this case, they are the result of least-squares, curve-fitting analysis of the satellite's actual orbit. Therefore, like the broadcast clock correction, accuracy of the broadcast ephemeris deteriorates with time. As a result, one of the most important parts of this portion of the Navigation message is called *AODE.* AODE is an acronym that stands for Age of Data Ephemeris, and it appears in both subframe 2 and 3.

Atmospheric Correction

Subframe 4 addresses atmospheric correction. As with subframe 1, the data offer only a partial solution to a problem. The Control Segment's monitoring stations find the apparent delay of a GPS signal caused by its trip through the ionosphere through an analysis of the different propagation rates of the two frequencies broadcast by all GPS satellites, L1 and L2. These two frequencies and the effects of the atmosphere on the GPS signal will be discussed later. For now, it is sufficient to say that a single-frequency receiver depends on the ionospheric correction in subframe 4 to help remove part of the error introduced by the atmosphere.

Antispoofing

Subframe 4 also contains a flag that tells the receiver when a security system, known as *antispoofing,* also known as simply AS, has been activated by the government ground control stations. Since December of 1993, the P code on all Block satellites has been encrypted to become the more secure Y code. Subframe 4 may also be asked to hold almanac information for satellites 25 through 32.

The Almanac

Subframe 5 tells the receiver where to find all the other GPS satellites. This subframe contains the ephemerides of up to 24 satellites. This is sometimes called *the almanac.* Here, the ephemerides are not complete. Their purpose is to help a GPS receiver lock onto more signals. Once the receiver finds its first satellite, it can look at the truncated ephemerides in subframe 5 of its Navigation message to figure the position of more satellites to track. But to collect any particular satellite's entire ephemeris, a receiver must acquire that satellite's signal and look there for subframes 2 and 3.

Satellite Health

Subframe 5 also includes *health data* for each satellite. GPS satellites are vulnerable to a wide variety of breakdowns, particularly clock trouble. That is one reason some of them carry as many as four clocks. Health data are also periodically uploaded by the ground control. Subframe 5 informs users of any satellite malfunctions before they try to use a particular signal.

Each of these five subframes begins with the same two words: the telemetry word TLM and the handover word HOW. Unlike nearly everything else in the Navigation message, these two words are generated by the satellite itself.

The TLM word is designed to indicate the status of uploading from the Control Segment while it is in progress. The HOW contains a number called the Z count, an important number for a receiver trying to acquire the P code, one of the primary GPS codes. The Z count tells the receiver exactly where the satellite stands in the generation of this very complicated code.

The P and C/A Codes

Like the Navigation message, the P and C/A codes are designed to carry information from the GPS satellites to the receivers. They, too, are impressed on the L1 and L2 carrier waves by modulation. However, unlike the Navigation message, the P and C/A codes are not vehicles for broadcasting information that has been uploaded by the Control Segment. They carry the raw data from which GPS receivers derive their time and distance measurements.

PRN

The P and C/A codes are complicated; so complicated, in fact, that they appear to be nothing but noise at first. And even though they are known as *pseudorandom* noise, or *PRN codes,* actually, these codes have been carefully designed. They have to be. They must be capable of repetition and replication.

P Code

For example, the P code generated at a rate of 10.23 million bits per second is available on both L1 and L2. Each satellite repeats its portion of the P code every 7 days, and the entire code is renewed every 37 weeks. All GPS satellites broadcast their codes on the same two frequencies, L1 and L2 for now, though a new frequency *(L5)* may be implemented—more about that later. Still, a GPS receiver must somehow distinguish one satellite's transmission from another. One method used to facilitate this satellite identification is the assignment of one particular week of the 37-week-long P code to each satellite. For example, space vehicle 14 *(SV 14)* is so named because it broadcasts the fourteenth week of the P code.

C/A Code

The C/A code is generated at a rate of 1.023 million bits per second; that is, 10 times slower than the P code. Here, satellite identification is quite straightforward. Not only does each GPS satellite broadcast a completely unique C/A code on its L1 frequency (and on L1 alone), but also the C/A code is repeated every millisecond.

SPS and PPS

The C/A code is the vehicle for the *SPS* (Standard Positioning Service), which is used for most civilian surveying applications. The P code on the other hand provides the same service for *PPS* (Precise Positioning Service). The current idea of SPS and PPS was developed by the Department of Defense. SPS is designed to provide a minimum level of positioning capability that is considered consistent with national security (at least ± 100 m, 95 % of the time) when intentionally degraded through SA (Selective Availability). But here's an update.

The intentional dithering of the satellite clocks by the Department of Defense called Selective Availability, or SA, was instituted in 1989. The accuracy of the C/A point positioning was too good! As mentioned above, the accuracy was supposed to be ± 100 meters, horizontally, 95 % of the time with a vertical accuracy of about ± 175 meters. But in fact, it turned out that the C/A-code point positioning gave civilians access to accuracy of about ± 20 meters to ± 40 meters. That wasn't according to plan, so they degraded the satellite clocks' accuracy on the C/A code on purpose until ± 100 meters was all you could get. The good news is that this error source is gone now!

Selective Availability was switched off on May 2, 2000 by presidential order. The intentional degradation of the satellite clocks is a thing of the past. To tell you the truth, Selective Availability never did hinder the surveying application of GPS much anyway; more about that later. But don't think that the satellite clocks don't contribute error to GPS positioning any more—they do.

PPS is designed for higher positioning accuracy and was originally available only to users authorized by the Department of Defense. That has changed somewhat; more about that later too.

The Production of a Modulated Carrier Wave

Since all the codes mentioned come to a GPS receiver on a modulated carrier, it is important to understand how a modulated carrier is generated. The signal created by an EDM is a good example of a modulated carrier.

EDM Ranging

As mentioned earlier, an EDM only needs one frequency standard because its electromagnetic wave travels to a retroprism and is reflected back to its origination. The EDM is both the transmitter and the receiver of the signal. Therefore, in general terms, the instrument can take half the time elapsed between the moment of transmission and the moment of reception, multiply by the speed of light, and find the distance between the itself and the retroprism (Distance = Elapsed Time × Rate).

Illustrated in Figure 1.2, the fundamental elements of the calculation of the distance measured by an EDM, ρ, are the time elapsed between transmission and reception of the signal, Δt, and the speed of light, c.

$$\text{Distance} = \rho$$
$$\text{Elapsed Time} = \Delta t$$
$$\text{Rate} = c.$$

GPS Ranging

However, the one-way ranging used in GPS is more complicated. It requires the use of two clocks. The broadcast signals from the satellites are collected by the

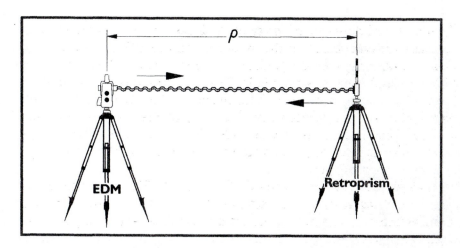

Figure 1.2. Two-Way Ranging.

receiver, not reflected. Nevertheless, in general terms, the full time elapsed between the instant a GPS signal leaves a satellite and arrives at a receiver, multiplied by the speed of light, is the distance between them.

Unlike the wave generated by an EDM, a GPS signal cannot be analyzed at its point of origin. The measurement of the elapsed time between the signal's transmission by the satellite and its arrival at the receiver requires two clocks, one in the satellite and one in the receiver. This complication is compounded because to correctly represent the distance between them these two clocks would need to be perfectly synchronized with one another. Since such perfect synchronization is physically impossible, the problem is addressed mathematically.

In Figure 1.3, the basis of the calculation of a range measured from a GPS receiver to the satellite, ρ, is the multiplication of the time elapsed between a signal's transmission and reception, Δt, by the speed of light, c.

A discrepancy of 1 microsecond from perfect synchronization, between the clock aboard the GPS satellite and the clock in the receiver can create a range error of 300 meters, far beyond the acceptable limits for nearly all surveying work.

Oscillators

The time measurement devices used in both EDM and GPS measurements are not really clocks. They are more correctly called oscillators, or *frequency standards*. In other words, they don't produce a steady series of ticks. They keep time by chopping a continuous beam of electromagnetic energy at extremely regular intervals. The result is a steady series of wavelengths and the foundation of the modulated carrier.

For example, the action of a shutter in a movie projector is analogous to the modulation of a coherent beam by the oscillator in an EDM. Consider the visible beam of light of a specific frequency passing through a movie projector. It is interrupted by the shutter, half of a metal disk rotating at a constant rate that alter-

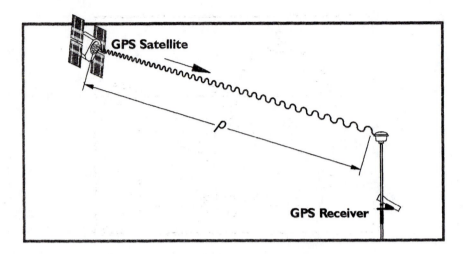

Figure 1.3. One-Way Ranging.

nately blocks and uncovers the light. In other words, the shutter chops the continuous beam into equal segments. Each length begins with the shutter closed and the light beam entirely blocked. As the shutter rotates open, the light beam is gradually uncovered. It increases to its maximum intensity, and then decreases again as the shutter gradually closes. The light is not simply turned on and off, it gradually increases and decreases. In this analogy the light beam is the carrier, and it has a wavelength much shorter than the wavelength of the modulation of that carrier produced by the shutter.

This modulation can be illustrated by a sine wave (Figure 1.4).

The wavelength begins when the light is blocked by the shutter. The first minimum is called a 0° *phase angle*. The first maximum is called the 90° phase angle and occurs when the shutter is entirely open. It returns to minimum at the 180° phase angle when the shutter closes again. But the wavelength isn't yet complete. It continues through a second shutter opening, 270°, and closing, 360°. The 360° phase angle marks the end of one wavelength and the beginning of the next one. The time and distance between every other minimum; that is, from the 0° to the 360° phase angles; is a wavelength and is usually symbolized by the Greek letter lambda, λ.

As long as the rate of an oscillator's operation is very stable, both the length and elapsed time between the beginning and end of every wavelength of the modulation will be the same.

A Chain of Electromagnetic Energy

GPS oscillators are sometimes called clocks because the frequency of a modulated carrier, measured in hertz, can indicate the elapsed time between the beginning and end of a wavelength, which is a useful bit of information for finding the distance covered by a wavelength. The length is approximately:

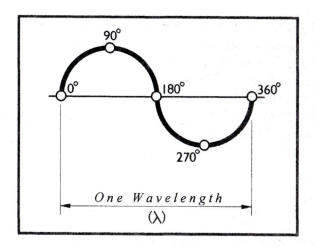

Figure 1.4. 0° to 360° = One Wavelength.

$$\lambda = \frac{c_a}{f}$$

Where: λ = the length of each complete wavelength in meters
c_a = the speed of light corrected for atmospheric effects
f = the frequency in hertz.

For example, if an EDM transmits a modulated carrier with a frequency of 9.84 MHz and the speed of light is approximately 300,000,000 meters per second (a more accurate value is 299,792,458 meters per second, but the approximation 300,000,000 meters per second will be used here for convenience), then:

$$\lambda = \frac{c_a}{f}$$

$$\lambda = \frac{300,000,000 \text{ mps}}{9,840,000 \text{ Hz}}$$

$$\lambda = 30.49 \text{ m}$$

the modulated wavelength would be about 30.49 meters long, or approximately 100 feet.

The modulated carrier transmitted from an EDM can be compared to a Gunter's chain constructed of electromagnetic energy instead of wire links. Each full link of this electromagnetic chain is a wavelength of a specific frequency. The measurement between an EDM and a reflector is doubled with this electronic chain because, after it extends from the EDM to the reflector, it bounces back to where it started. The entire trip represents twice the distance and is simply divided by 2. But like the surveyors who used the old Gunter's chain, one cannot depend that a particular measurement will end conveniently at the end of a complete link (or wavelength). A measurement is much more likely to end at some fractional part of a link (or wavelength). The question is, where?

Phase Shift

With the original Gunter's chain, the surveyor simply looked at the chain and estimated the fractional part of the last link that should be included in the measurement. Those links were tangible. Since the wavelengths of a modulated carrier are not, the EDM must find the fractional part of its measurement electronically. Therefore, it does a comparison. It compares the phase angle of the returning signal to that of a replica of the transmitted signal to determine the *phase shift*. That phase shift represents the fractional part of the measurement. This principle is used in distance measurement by both EDM and GPS systems.

How does it work? First, it is important to remember that points on a modulated carrier are defined by phase angles, such as $0°$, $90°$, $180°$, $270°$, etc. (Figure 1.4). When two modulated carrier waves reach exactly the same phase angle at exactly the same time, they are said to be *in phase, coherent,* or *phase locked.* However, when two waves reach the same phase angle at different times, they are *out of phase* or *phase shifted.* For example, in Figure 1.5, the sine wave shown in the gray dashed line has returned to an EDM from a reflector. Compared with the sine wave shown in the dark solid line, it is out of phase by one-quarter of a wavelength. The distance between the EDM and the reflector, ρ, is then:

$$\rho = \frac{(n\lambda + d)}{2}$$

where: n = the number of full wavelengths the modulated carrier has completed
         d = the fractional part of a wavelength at the end that completes the doubled distance.

In this example, d is three-quarters of a wavelength because it lacks its last quarter. But how would the EDM know that?

It knows because at the same time an external carrier wave is sent to the reflector, the EDM keeps an identical internal reference wave at home in its receiver circuits. In Figure 1.6 the external beam returned from the reflector is compared to the reference wave and the difference in phase between the two can be measured.

Both EDM and GPS ranging use the method represented in this illustration. In GPS, the measurement of the difference in the phase of the incoming signal and the phase of the internal oscillator in the receiver reveals the small distance at the

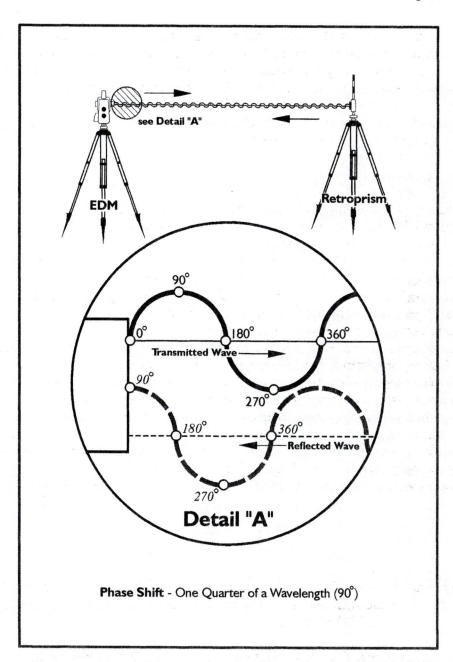

Phase Shift - One Quarter of a Wavelength (90°)

Figure 1.5. An EDM Measurement.

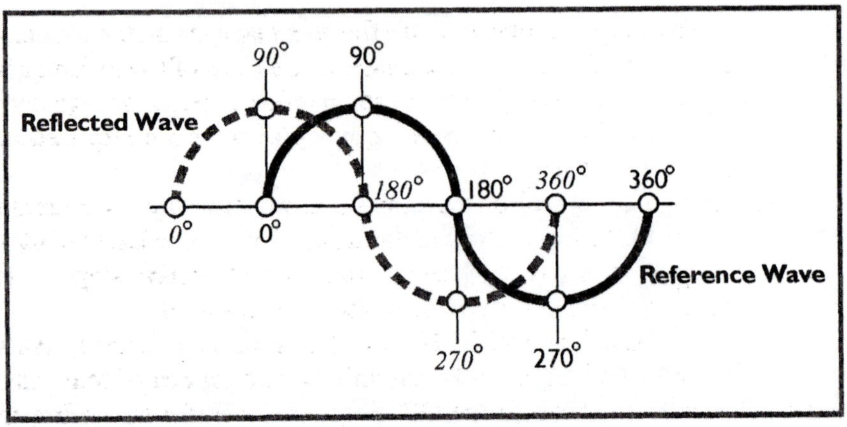

Figure 1.6. Reference and Reflected Waves.

end of a range. In GPS, the process is called *carrier phase ranging*. And as the name implies, this measurement is actually done on the <u>carrier</u> itself.

The Cycle Ambiguity Problem

While this technique discloses the fractional part of a wavelength, a problem remains—determining the number of full wavelengths of the EDM's modulated carrier between the transmitter and the receiver. This cycle ambiguity problem is solved in EDM measurements by modulating the carrier in ever longer wavelengths. For example, the meter and part of a meter aspects of a measured distance can be resolved by measuring the phase difference of a 10 meter wavelength. This procedure may be followed by the resolution of the tens of meters using a wavelength of 100 meters. The hundreds of meters can then be resolved with a wavelength of 1,000 meters, and so on. Actually three wavelengths, 10 meters, 1,000 meters, and 10,000 meters, are used in most EDMs.

Such a method is convenient for the EDM's two-way ranging system, but impossible in the one-way ranging used in GPS measurements. GPS ranging must use an entirely different strategy for solving the cycle ambiguity problem because the satellites broadcast only two carriers of constant wavelengths, in one direction, from the satellites to the receivers. Unlike an EDM measurement, the wavelengths of these carriers in GPS cannot be changed to resolve the number of cycles between transmission and reception. Still, the carrier phase measurements remain an important *observable* in GPS ranging.

TWO OBSERVABLES

The word *observable* is used throughout GPS literature to indicate the signals whose measurement yields the range or distance between the satellite and the receiver. The word is used to draw a distinction between the thing being measured, the observable, and the measurement, the observation.

In GPS there are two types of observables: the *pseudorange* and the carrier phase. The latter, also known as the *carrier beat phase,* is the basis of the techniques used for high-precision GPS surveys. On the other hand, the pseudorange can serve applications when virtually instantaneous point positions are required or relatively low accuracy will suffice.

These basic observables can also be combined in various ways to generate additional measurements that have certain advantages. It is in this latter context that pseudoranges are used in many GPS receivers as a preliminary step toward the final determination of position by carrier phase measurement.

The foundation of pseudoranges is the correlation of code carried on a modulated carrier wave received from a GPS satellite with a replica of that same code generated in the receiver. Most of the GPS receivers used for surveying applications are capable of code correlation. That is, they can determine pseudoranges from the C/A code or the P code. These same receivers are usually capable of determining ranges using the unmodulated carrier as well. However, first let us concentrate on the pseudorange.

Encoding by Phase Modulation

A carrier wave can be modulated in various ways. Radio stations use carrier waves that are AM, amplitude modulated or FM, frequency modulated. When your radio is tuned to 105 FM, you are not actually listening to 105 MHz despite the announcer's assurances—it is well above the range of human hearing. 105 MHz is just the frequency of the carrier wave that is being modulated. Those modulations occur at a much slower rate, making the speech and music intelligible.

The GPS carriers L1 and L2 could have been modulated in a variety of ways too, but the C/A and P codes impressed on the GPS carriers are the result of phase modulations. One consequence of this method of modulation is that the signal can occupy a broader bandwidth than it otherwise would. The GPS signal is said to have a *spread spectrum* because of its intentionally increased bandwidth. This characteristic offers several advantages, including more accurate ranging, increased security, and less interference.

The particular kind of phase modulation used to encode a carrier in GPS is known as *binary biphase* modulation, in which each zero and one of the binary code is known as a *code chip.* Zero represents the *normal* state, and one represents the *mirror image* state. In other words, the frequency and amplitude of the carrier remain constant, they are not modulated. The modulations from zero to one and from one to zero are accomplished by instantaneous 180° changes in phase. Note in Figure 1.7 that each shift from zero to one, and from one to zero, is accompanied by a corresponding change in the phase of the carrier.

The rate of all of the components of GPS signals are multiples of the standard rate of the oscillators, 10.23 MHz. This rate is known as the *fundamental clock rate* and is symbolized F_o. For example, the GPS carriers are 154 times F_o, or 1575.42 MHz, and 120 times F_o, or 1227.60 MHz, L1 and L2, respectively.

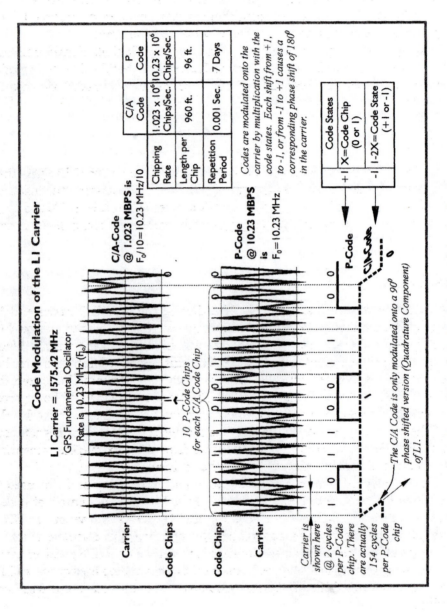

Figure 1.7. Code Modulation of the L1 Carrier.

The codes are also based on F_o. 10.23 *code chips* of the P code, zeros or ones, occur every microsecond. In other words, the *chipping rate* of the P code is 10.23 million bits per second, 10.23 Mbps, exactly the same as F_o, 10.23 MHz.

The chipping rate of the C/A code is 10 times slower than the P code, a tenth of F_o, 1.023 Mbps. Ten P code chips occur in the time it takes to generate one C/A code chip, allowing P code derived pseudoranges to be much more precise. This is one reason the C/A code is known as the *coarse/acquisition code*.

Even though both codes are broadcast on L1, they are distinguishable from one another by their transmission in *quadrature*. That means that the C/A code modulation on the L1 carrier is phase shifted 90° from the P code modulation on the same carrier.

PSEUDORANGING

Strictly speaking, a pseudorange observable is based on a time shift. This time shift can be symbolized by $d\tau$, d tau, and is the time elapsed between the instant a GPS signal leaves a satellite and the instant it arrives at a receiver. The concept can be illustrated by the process of setting a watch from a time signal heard over a telephone.

Propagation Delay

Imagine that a recorded voice said, "The time at the tone is 3 hours and 59 minutes." If a watch was set at the instant the tone was <u>heard</u>, the watch would be wrong. Supposing that the moment the tone was broadcast was indeed 3 hours and 59 minutes; the moment the tone is heard must be a bit later. It is later because of the time it took the tone to travel through the telephone lines from the point of broadcast to the point of reception. This elapsed time would be approximately equal to the length of the circuitry traveled by the tone divided by the speed of the electricity, which is the same as the speed of all electromagnetic energy, including light and radio signals. In fact, it is possible to imagine measuring the actual length of that circuitry by doing the division.

In GPS, that elapsed time is known as the *propagation* delay and it is used to measure length. The measurement is accomplished by a combination of codes. The idea is somewhat similar to the strategy used in EDMs. But where an EDM generates an internal replica of its <u>modulated carrier wave</u> to correlate with the signal it receives by reflection, a pseudorange is measured by a GPS receiver using a replica of the <u>code</u> that has been impressed on the modulated carrier wave. The GPS receiver generates this replica itself to compare with the code it receives from the satellite.

Code Correlation

To conceptualize the process, one can imagine two codes generated at precisely the same time and identical in every regard: one in the satellite and one in the receiver. The satellite sends its code to the receiver but on its arrival, the two codes don't line up. The codes are identical but they don't *correlate* until the replica code in the receiver is time shifted.

Then the receiver generated replica code is shifted relative to the received satellite code. It is this time shift that reveals the propagation delay, the time it took the signal to make the trip from the satellite to the receiver, $d\tau$. It is the same idea described above as the time it took the tone to travel through the telephone lines, except the GPS code is traveling through space and atmosphere. Once the time shift is accomplished, the two codes match perfectly and the time the satellite signal spent in transit has been measured; well, almost.

It would be wonderful if that time shift could simply be divided by the speed of light and yield the true distance between the satellite and the receiver at that instant, and it is close. But there are physical limitations on the process that prevent such a perfect relationship. More about that later.

Autocorrelation

Actually lining up the code from the satellite with the replica in the GPS receiver is called *autocorrelation*, and depends on the transformation of code chips into *code states*. The formula used to derive code states ($+1$ and -1) from code chips (0 and 1) is:

$$\text{code state} = 1 - 2x$$

where x is the code chip value. For example, a normal code state is $+1$, and corresponds to a code chip value of 0. A mirror code state is -1, and corresponds to a code chip value of 1.

The function of these code states can be illustrated by asking two questions:

First, if a tracking loop of 10 code states generated in a receiver does <u>not</u> match 10 code states received from the satellite, how does the receiver know? In that case, the sum of the products of each of the receiver's 10 code states, with each of the 10 from the satellite, when divided by 10, *does not equal 1*.

Secondly, how does the receiver know when a tracking loop of 10 replica code states <u>does</u> match 10 code states from the satellite? In that case, the sum of the products of each code state of the receiver's replica 10, with each of the 10 from the satellite, divided by 10, *is exactly 1*.

The *autocorrelation function* is:

$$\frac{1}{N}\int_0^T X(t) * X(t-\tau)\,dt = \frac{1}{N}\sum_{i=1}^{N} X_i * X_{i-j}$$

In Figure 1.8, before the code from the satellite and the replica from the receiver are matched:

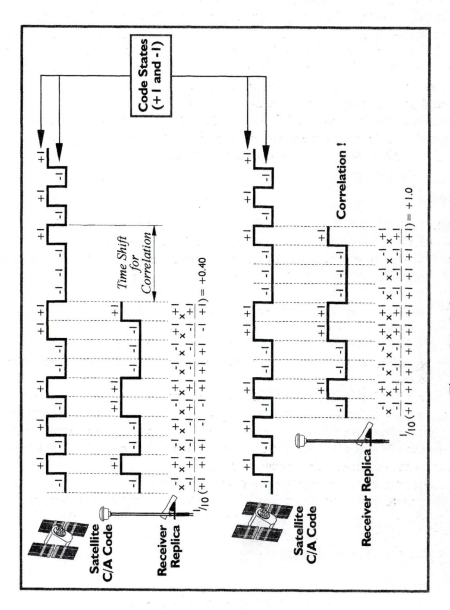

Figure 1.8. Code Correlation.

$$\frac{1}{10}\sum_{i=1}^{10} X_i * X_i = \frac{1}{10}(+1+1+1-1-1+1+1+1-1+1) = +0.40$$

the sum of the products of the code states is not 1.

Following the correlation of the two codes:

$$\frac{1}{10}\sum_{i=1}^{10} X_i * X_i = \frac{1}{10}(+1+1+1+1+1+1+1+1+1+1) = +1.0$$

the sum of the code states is exactly 1, and the receiver's replica code fits the code from the satellite like a key fits a lock.

Lock and the Time Shift

Once correlation of the two codes is achieved with a *delay lock loop,* it is maintained by a *correlation channel* within the GPS receiver, and the receiver is sometimes said to have *achieved lock,* or to be *locked on* to the satellites. If the correlation is somehow interrupted later, the receiver is said to have *lost lock.* However, as long as the lock is present the Navigation message is available to the receiver. Remember that one of its elements is the broadcast clock correction that relates the satellite's onboard clock to GPS time, and a limitation of the pseudorange process comes up.

Imperfect Oscillators

One reason the time shift, $d\tau$, found in autocorrelation cannot quite reveal the true range, ρ, of the satellite at a particular instant is the lack of perfect synchronization between the clock in the satellite and the clock in the receiver. Please recall that the two compared codes are generated directly from the fundamental rate, F_o, of those clocks. And since these widely separated clocks, one on earth and one in space, cannot be in perfect lockstep with one another the codes they generate cannot be in perfect synch either. Therefore, a small part of the observed time shift, $d\tau$, must always be due to the disagreement between these two clocks. In other words, the time shift not only contains the signal's transit time from the satellite to the receiver, it contains clock errors too.

In fact, whenever satellite clocks and receiver clocks are checked against the carefully controlled GPS time they are found to be drifting a bit. Their oscillators are imperfect. It is not surprising that they are not quite as stable as the more than 150 atomic clocks around the world that are used to define the rate of GPS time. They are subject to the destabilizing effects of temperature, acceleration, radiation, and other inconsistencies. As a result, there are two clock offsets that bias every satellite to receiver pseudorange observable. That is one reason it is called a *pseudo*range.

A Pseudorange Equation

Clock offsets are only one of the errors in pseudoranges. Their relationship can be illustrated by the following equation (Langley, 1993):

$$p = \rho + c\left(dt - dT\right) + d_{ion} + d_{trop} + \varepsilon_p$$

where: p = the pseudorange measurement
ρ = the true range
c = the speed of light
dt = the satellite clock offset from GPS time
dT = the receiver clock offset from GPS time
d_{ion} = ionospheric delay
d_{trop} = tropospheric delay
ε_p = multipath, receiver noise, etc.

Please note that the pseudorange, p, and the true range, ρ, cannot be made equivalent, without consideration of clock offsets, atmospheric effects, and other biases that are inevitably present.

This discussion of time can make it easy to lose sight of the real objective, which is the position of the receiver. Obviously, if the coordinates of the satellite and the coordinates of the receiver were known perfectly, then it would be a simple matter to determine time shift dt or find the true range ρ between them.

In fact, receivers placed at known coordinated positions can establish time so precisely they are used to monitor atomic clocks around the world. Several receivers simultaneously tracking the same satellites can achieve resolutions of 10 nanoseconds or better. Also, receivers placed at known positions can be used as base stations to establish the relative position of receivers at unknown stations, a fundamental principle of most GPS surveying.

It can be useful to imagine the true range term ρ, also known as the *geometric range*, actually includes the coordinates of both the satellite and the receiver. However, they are hidden within the measured value, the pseudorange, p, and all of the other terms on the right side of the equation. The objective then is to mathematically separate and quantify these biases so that the receiver coordinates can be revealed. Clearly, any deficiency in describing, or *modeling*, the biases will degrade the quality of the final determination of the receiver's position (see Figure 1.9).

The One-Percent Rule of Thumb

Here is convenient approximation. The maximum resolution available in a pseudorange is about 1 percent of the chipping rate of the code used, whether it is the P code or the C/A code. In practice, positions derived from these codes are rather less reliable than described by this approximation, but it offers a basis to compare P code and C/A code pseudoranging.

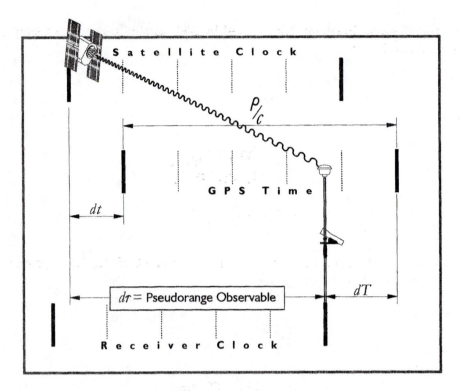

Figure 1.9. Pseudorange.

A P-code chip occurs every 0.0978 of a microsecond. Therefore, a P-code based measurement can have a maximum precision of about 1 percent of about a tenth of a microsecond, or one nanosecond. One nanosecond multiplied by the speed of light is approximately 30 centimeters, 1 percent of the length of a single P-code chip.

The C/A-code based pseudorange is 10 times less precise. Its chipping rate is 10 times slower. A C/A-code pseudorange has a maximum resolution of about 3 meters; that is, 1 percent of the length of a single C/A-code chip.

This same 1 percent rule of thumb can illustrate the increased precision of the carrier phase observable over the pseudorange. First, the length of a single wavelength of each carrier is calculated using the same formula as was used previously.

$$\lambda = \frac{c_a}{f}$$

Where: λ = the length of each complete wavelength in meters
c_a = the speed of light corrected for atmospheric effects
f = the frequency in hertz.

The L1-1575.42 MHz *carrier* transmitted by GPS satellites has a wavelength of approximately 19 cm.

$$\lambda = \frac{c_a}{f}$$

$$\lambda = \frac{300 \times 10^6 \text{ mps}}{1575.42 \times 10^6 \text{ Hz}}$$

$$\lambda = 0.19 \text{ m}$$

The L2-1227.60 MHz frequency *carrier* transmitted by GPS satellites has a wavelength of approximately 24 cm.

$$\lambda = \frac{c_a}{f}$$

$$\lambda = \frac{300 \times 10^6 \text{ mps}}{1227.60 \times 10^6 \text{ Hz}}$$

$$\lambda = 0.24 \text{ m}$$

Therefore, using either carrier, the carrier phase measurement resolved to 1 percent of the wavelength would be about 2 mm.

A 3 m ranging precision is not adequate for most land surveying applications, not to say all. However, the actual positional accuracy of a single C/A code receiver was about ±100 m with selective availability, SA, turned on and ±30 m with SA turned off unless differential, *DGPS,* techniques are used. More about that later. But carrier phase observations are certainly the preferred method for the higher precision work most surveyors have come to expect from GPS.

CARRIER PHASE RANGING

Carrier Phase Comparisons

Carrier phase is the observable at the center of high accuracy surveying applications of GPS. It depends on the carrier waves themselves, the unmodulated L1 and L2, rather than their P and C/A codes. One bit of good news was always apparent from this strategy right away—the user was immune from SA, selective availability. As mentioned earlier, SA was the intentional degradation of the SPS, the standard positioning service available through the C/A code. Since carrier phase observations do not use codes, they were never affected by SA. Now that SA has been turned off, the point is moot, unless SA should be reinstituted.

Understanding carrier phase is perhaps a bit more difficult than the pseudorange, but the basis of the measurements has some similarities. For example, the foundation of a pseudorange measurement is the correlation of the codes received from a GPS satellite with replicas of those codes generated within the receiver. The foundation of the carrier phase measurement is the combination of the unmodulated carrier itself received from a GPS satellite with a replica of that carrier generated within the receiver.

Phase Difference

A few similarities can also be found between a carrier phase observation and a distance measurement by an EDM. As mentioned earlier, an EDM sends a modulated carrier wave to the reflector, and generates an identical internal reference. When the external beam returns from the reflector, it is compared with the reference wave. The difference in phase between the two reveals the fractional part of the measurement, even though the number of complete cycles between the EDM and the reflector may not be immediately apparent until modulated carriers of longer wavelengths are used.

Likewise, it is the phase difference between the incoming signal and the internal reference that reveals the fractional part of the carrier phase measurement in GPS. The incoming signal is from a satellite rather than a reflector of course, but like an EDM measurement, the internal reference is derived from the receiver's oscillator and the number of complete cycles is not immediately known.

Beat

The carrier phase observable is sometimes called the *reconstructed carrier phase* or *carrier beat phase* observable. In this context, a *beat* is the pulsation resulting from the combination of two waves with different frequencies (Figure 1.10). An analogous situation occurs when two musical notes of different pitch are sounded at the same time. Their two frequencies combine and create a third note, called the beat. Musicians can tune their instruments by listening for the beat that occurs when two pitches differ slightly. This third pulsation may have a frequency equal to the difference or the sum of the two original frequencies.

The beat phenomenon is by no means unique to musical notes. It can occur when any pair of oscillations with different frequencies are combined. In GPS a beat is created when a carrier generated in a GPS receiver and a carrier received from a satellite are combined. At first, that might not seem sensible.

How could a beat be created by combining two absolutely identical unmodulated carriers? There should be no difference in frequency between an L1 carrier generated in a satellite and an L1 carrier generated in a receiver. They both should have a frequency of 1575.42 MHz. If there is no difference in the frequencies, how can

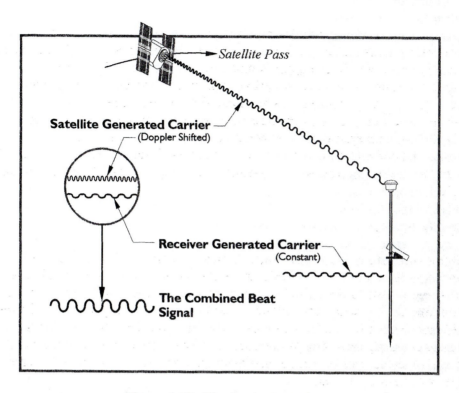

Figure 1.10. The Carrier Beat Phase.

there be a beat? But there is a slight difference between the two carriers. Something happens to the frequency of the carrier on its trip from a GPS satellite to a receiver. Its frequency changes. The phenomenon is described as the Doppler effect.

The Doppler Effect

Sound provides a model for the explanation of the phenomenon. An increase in the frequency of a sound is indicated by a rising pitch; a lower pitch is the result of a decrease in the frequency. A stationary observer listening to the blasting horn on a passing train will note that as the train gets closer, the pitch rises, and as the train travels away, the pitch falls. Furthermore, the change in the sound, clear to the observer standing beside the track, is not heard by the engineer driving the train. He hears only one constant, steady pitch. The relative motion of the train with respect to the observer causes the apparent variation in the frequency of the sound of the horn.

In 1842, Christian Doppler described this frequency shift now named for him. He used the analogy of a ship on an ocean with equally spaced waves. In his allegory, when the ship is stationary the waves strike it steadily, one each second. But if the ship sails into the waves, they break across its bow more frequently. If the ship then turns around and sails with the waves, they strike less frequently across its stern. The waves themselves have not changed; their frequency is constant. But to the observer on the ship their frequency seems to depend on his motion.

GPS and the Doppler Effect

From the observer's point of view, whether it is the source, the observer, or both that are moving, the frequency increases while they move together and decreases while they move apart. Therefore, if a GPS satellite is moving toward an observer, its carrier wave comes into the receiver with a higher frequency than it had at the satellite. If a GPS satellite is moving away from the observer, its carrier wave comes into the receiver with a lower frequency than it had at the satellite. Since a GPS satellite is virtually always moving with respect to the observer, any signal received from a GPS satellite is Doppler-shifted.

A Carrier Phase Approximation

The carrier phase observation in cycles is often symbolized by *phi*, ϕ, in GPS literature. Other conventions include the use of superscripts to indicate satellite designations and the use of subscripts to define receivers. For example, in the following equation ϕ_r^s is used to symbolize the carrier phase observation between satellite *s* and receiver *r*. The difference that defines the carrier beat phase observation is (Wells et al., 1986):

$$\phi = \phi_r^s = \phi^s(t) - \phi_r(T)$$

where $\phi^s(t)$ is the phase of the carrier broadcast from the satellite *s* at time *t*. Please note that the frequency of this carrier is the same, nominally constant frequency, that is generated by the receiver's oscillator. $\phi_r(T)$ is its phase when it reaches the receiver *r* at time *T*.

A description of the use of the carrier phase observable to measure range can start with the same basis as the calculation of the pseudorange, travel time. The time elapsed between the moment the signal is broadcast, *t*, and the moment it is received, *T*, multiplied by the speed of light, *c*, will yield the range between the satellite and receiver, ρ:

$$(T - t)c \approx \rho$$

Even using this simplified equation it is possible to get an approximate idea of the relation between time and range for some assumed, nominal values. These values could not be the basis of any actual carrier phase observation because, among other reasons, it is not possible for the receiver to know when a particular carrier wave left the satellite. But for the purpose of illustration, suppose a carrier left a satellite at 00 hr 00 min 00.000 sec, and arrived at the receiver 67 milliseconds later.

$$(T-t)c \approx \rho$$

$$(00\text{h}:00\text{m}:00.067\text{s}-00\text{h}:00\text{m}:00.000\text{s})300{,}000 \text{ km/s} \approx \rho$$

$$(0.067)300{,}000 \text{ km/s} \approx \rho$$

$$20{,}100 \text{ km} \approx \rho$$

This estimate indicates that if the carrier broadcast from the satellite reaches the receiver 67 milliseconds later, the range between them is approximately 20,100 km.

Carrying this example a bit farther, the wavelength, λ, of the L1 carrier can be calculated by dividing the speed of light, c, by the L1 frequency, f:

$$\lambda = \frac{c}{f}$$

$$\lambda = \frac{300{,}000{,}000 \text{ m/s}}{1{,}575{,}420{,}000 \text{ Hz}}$$

$$\lambda = 0.1904254104 \text{ m}$$

Dividing the approximated range, ρ, by the calculated L1 carrier wavelength, λ, yields a rough estimation of the carrier phase in cycles ϕ:

$$\frac{\rho}{\lambda} \approx \phi$$

$$\frac{20{,}100{,}000 \text{ m}}{0.1904254104 \text{ m}} \approx \phi$$

$$105{,}553{,}140 \text{ cycles} \approx \phi$$

The 20,100 km range implies that the L1 carrier would cycle through approximately 105,553,140 wavelengths on its trip from the satellite to the receiver.

Of course, these relationships are much simplified. However, they can be made fundamentally correct by recognizing that ranging with the carrier phase observable is subject to all of the same biases and errors as the pseudorange. For example, terms such as the receiver clock offset may be incorporated, again symbolized by dT as it was in the pseudorange equation. The imperfect satellite clock can be included, its error is symbolized by dt. The tropospheric delay d_{trop}, the ionospheric delay d_{ion}, and multipath-receiver noise ε_ϕ, are also added to the range measurement. The ionospheric delay will be negative here; more about that later. With these changes the simplified travel time equation can be made a bit more realistic.

$$\left[(T+dT)-(t+dt)\right]c = \rho - d_{ion} + d_{trop} + \varepsilon_\phi$$

This, more realistic, equation can be rearranged to isolate the elapsed time $(T-t)$ on one side, by dividing both sides by c, and then moving the clock errors to the right side (Wells et al., 1986).

$$\frac{\left[(T+dT)-(t+dt)\right]c}{c} = \frac{\rho - d_{ion} + \varepsilon_\phi}{c}$$

$$\left[(T+dT)-(t+dt)\right] = \frac{\rho - d_{ion} + d_{trop} + \varepsilon_\phi}{c}$$

$$(T-t+dT-dt) = \frac{\rho - d_{ion} + d_{trop} + \varepsilon_\phi}{c}$$

$$(T-t+DT-dt) = \frac{\rho - d_{ion} + d_{trop} + \varepsilon_\phi}{c} + (dt-dT)$$

$$T-t = dt - dT + \frac{\rho - d_{ion} + d_{trop} + \varepsilon_\phi}{c}$$

This expression now relates the travel time to the range. However, in fact, a carrier phase observation cannot rely on the travel time, for two reasons. First, in a carrier phase observation the receiver has no codes with which to tag any particular instant on the incoming continuous carrier wave. Second, since the receiver cannot distinguish one cycle of the carrier from any other, it has no way of knowing the initial phase of the signal when it left the satellite. In other words, the receiver cannot know the travel time and, therefore, it is hard to see how it can determine the number of complete cycles between the satellite and itself. This unknown quantity is called the *cycle ambiguity*.

Remember, the approximation of 105,553,140 wavelengths in this example is for the purpose of comparison and illustration only. In actual practice, a carrier phase observation must derive the range from a measurement of phase at the receiver, not from a known travel time of the signal.

The missing information is the number of complete phase cycles between the receiver and the satellite at the instant that the tracking began. The critical unknown integer, symbolized by N, is the cycle ambiguity, and it cannot be directly measured by the receiver. The receiver can count the complete phase cycles it receives from the moment it starts tracking until the moment it stops. It can also monitor the fractional phase cycles, but the cycle ambiguity N is unknown.

An Illustration of the Cycle Ambiguity Problem

The situation is somewhat analogous to an unofficial technique used by some nineteenth century contract surveyors on the Great Plains. The procedure can be used as a rough illustration of the cycle ambiguity problem in GPS.

It was known as the *buggy wheel* method of chaining. Some of the lines of the public land system that crossed open prairies were originally surveyed by loading a wagon with stones or stakes and tying a cloth to a spoke of the wheel. One man drove the team, another kept the wagon on line with a compass, and a third counted the revolutions of the flagged wheel to measure the distance. When there had been enough turns of the improvised odometer to measure half a mile they set a stone or stake to mark the corner and then rolled on, counting their way to the next corner.

A GPS receiver is like the man assigned to count the turns of the wheel. He is supposed to begin his count from the moment the crew leaves the newly set corner, but instead, suppose he jumps into the wagon, gets comfortable and takes an unscheduled nap. When he wakes up the wagon is on the move. Trying to make up for his laxness, he immediately begins counting. But at that moment the wheel is at a half turn, a fractional part of a cycle. He counts the subsequent half turn and then, back on the job, he intently counts each and every full revolution as they come around. His tally grows as the cycles accumulate, but he is in trouble and he knows it. He cannot tell how far the wagon has traveled; he was asleep for the first part of the trip. He has no way of knowing how far they had come before he woke up and started counting. He is like a GPS receiver that cannot know how far it is from the satellite when it starts counting phase cycles. They can tell it nothing about how many cycles stood between itself and the satellite when the receiver was locked on and began tracking. Those unknown cycles are the cycle ambiguity.

The 360° cycles in the carrier phase observable are wavelengths λ, not revolutions of a wheel. Therefore, the cycle ambiguity included in the complete carrier phase equation is an integer number of wavelengths, symbolized by λN (Langley, 1993). So, the complete carrier phase observable equation can now be stated as:

$$\Phi = \rho + c\left(dt - dT\right) + \lambda N - d_{ion} + d_{trop} + \varepsilon_\Phi$$

EXERCISES

1. What is the function of the information in subframe 5 of the Navigation message?

   (a) Once a receiver is locked onto a satellite it helps the receiver to determine the position of the satellite that is transmitting the Navigation message.
   (b) Once a receiver is locked onto a satellite it helps the receiver to correct the part of the delay of the signal caused by the ionosphere.
   (c) Once a receiver is locked onto a satellite it helps the receiver to correct the received satellite time to GPS time.
   (d) Once a receiver is locked onto a satellite it helps the receiver to acquire the signals of the other satellites.

2. Which height most correctly expresses a nominal altitude of GPS satellites above the earth?

   (a) 20,000 miles
   (b) 35,420 km
   (c) 20,183,000 m
   (d) 108,000 nautical miles

3. Which of the following statements about the clocks in GPS satellites is not correct?

   (a) The time signal for each satellite is independent from the other satellites and is generated from its own onboard clock.
   (b) The clocks in GPS satellites may also be called oscillators or frequency standards.
   (c) Every GPS satellite is launched with very stable atomic clocks onboard.
   (d) The clocks in any one satellite are allowed to drift up to one nanosecond from GPS time before it is tweaked by the Control Segment.

4. The Global Positioning System is known as a passive system. What does that mean?

   (a) The ranges are measured with signals in the microwave part of the electromagnetic spectrum.
   (b) Only the satellites transmit signals; the users receive them.
   (c) A GPS receiver must be able to gather all the information it needs to determine its own position from the signals it bounces off the satellites.
   (d) The signals from a GPS receiver return to the satellite.

5. Which comparison of EDM and GPS processes is correct?

   (a) EDM and GPS signals are both reflected back to their sources.
   (b) EDM measurements require atmospheric correction, GPS ranges do not.
   (c) EDMs and GPS satellites both transmit modulated carriers.
   (d) Phase differencing is used in EDM measurement, but not in GPS.

6. What information below is critical to defining the relationship between GPS time and UTC?

   (a) multipath
   (b) the broadcast ephemeris
   (c) an antispoofing flag
   (d) leap seconds

7. The P code and the C/A code are created by shifts from 0 to 1 or 1 to 0 known as code chips. There is a corresponding change of 180° in the GPS carrier waves. What changes?

   (a) the fundamental clock rate
   (b) the frequency
   (c) the phase
   (d) the amplitude

8. The P code and the C/A code are both broadcast on L1. How can a GPS receiver distinguish between them on that carrier?

   (a) the chipping rate for the C/A code is 10 times slower than the chipping rate for the P code.
   (b) they are broadcast in binary biphase modulation.
   (c) the fundamental clock rate is 10.23 MHz.
   (d) they are broadcast in quadrature.

9. Which of the following is the most correct description of the Doppler effect?

   (a) the distortion of electromagnetic waves due to the density of charged particles in the earth's upper atmosphere
   (b) the systematic changes that occur when a moving object or light beam pass through a gravitational field
   (c) the systematic changes that occur when a moving object approaches the speed of light
   (d) the shift in frequency of an acoustic or electromagnetic radiation emitted by a source moving relative to an observer

10. When Selective Availability (SA) was switched off, how did it affect surveying applications of GPS?

    (a) the accuracy of most surveying applications of GPS doubled.
    (b) the P code was not encrypted.
    (c) the cycle ambiguity problem was eliminated.
    (d) there was no significant change.

ANSWERS AND EXPLANATIONS

1. Answer is (d)

   Explanation: Subframe 5 contains the ephemerides of up to 24 satellites. The purpose of subframe 5 is to help the receiver acquire the signals of the other satellites. The receiver must first lock onto a satellite to have access to the Navigation message, of course, but once the Navigation message can be read, the positions of all of the other satellites can be computed.

2. Answer is (c)

   Explanation: The GPS constellation is still evolving, but the orbital configuration is fairly well settled. Each GPS satellite's orbit is nearly circular and has a nominal altitude of 20,183 km, or 20,183,000 meters.
   The orbit is approximately 12,500 statute miles or 10,900 nautical miles above the surface of the planet. GPS satellites are higher than the usual orbit assigned to most other satellites, including the space shuttle. Their high altitude allows each GPS satellite to be viewed simultaneously from a large portion of the earth at any given moment. However, GPS satellites are well below the height, 35,420 km, required for the sort of geosynchronous orbit used for communications satellites.

3. Answer is (d)

   Explanation: Government tracking facilities monitor the drift of each satellite clock's deviation from GPS time. The drift is allowed to reach a maximum of one millisecond before it is adjusted. Until the difference reaches that level, it is contained in the broadcast clock correction in subframe 1 of the Navigation message.

4. Answer is (b)

   Explanation: Some of the forerunners of GPS were navigational systems that involved transmissions from the users, but not GPS. The military designed the system to exclude anything that would reveal the location of the GPS receiver. If GPS had been built as a two-way system it would have been much more complicated, especially as the number of users grew. Therefore, it is a passive system, meaning that the satellites transmit and the users receive.

5. Answer is (c)

Explanation: An unmodulated carrier carries no information and no code. While GPS receivers can and do make phase measurements on the unmodulated carrier waves they also make use of the code information available on the modulated carrier transmitted by GPS satellites to determine pseudoranges. GPS uses a one-way system. The modulated carrier with code information travels from the satellite to the receiver, where it is correlated with a reference. This one-way method requires a frequency standard in both the satellite and the receiver.

An EDM's measurement is also based on a modulated carrier. However, the EDM uses a two-way system. The modulated carrier is transmitted to a retroprism. It is reflected and returns to the EDM. The EDM can then determine the phase delay by comparing the returned modulated wave with a reference in the instrument. This two-way method requires only one frequency standard in the EDM since the modulated carrier is reflected.

6. Answer is (d)

Explanation: GPS time is calculated using the atomic clock at the Master Control Station (MCS) at the Schriever Air Force Base, formerly known as Falcon Air Force Station, near Colorado Springs, Colorado. GPS time is kept within 1 millisecond of UTC. However, UTC is adjusted for leap seconds, and GPS time is not. Conversion of GPS time to UTC requires knowledge of the leap seconds applied to UTC since January 1980. This information is available from the United States Naval Observatory, *USNO,* time announcements.

7. Answer is (c)

Explanation: Phase modulations are used to mark the divisions between the code chips. Whenever the C/A code or the P code switches from a binary 1 to a binary 0 or vice versa, its L1 or L2 carriers have a sharp mirror-image shift in phase. About every millionth of a second the phase of the C/A code carrier can shift. But the P code carrier can have a phase shift about every ten millionth of a second.

8. Answer is (d)

Explanation: Even though both codes are broadcast on L1, they are distinguishable from one another by their transmission in *quadrature.* That means that the C/A code modulation on the L1 carrier is phase shifted 90° from the P code modulation on the same carrier.

9. Answer is (d)

Explanation: If the source of radiation is moving relative to an observer, there is a difference between the frequency perceived by the observer and the frequency of the radiation at its source. The shift is to higher frequencies when the source moves toward the observer and to lower frequencies when it moves away. This effect has been named for its discoverer, C.J. Doppler.

10. Answer is (d)

Explanation: When Selective Availability (SA) was switched off on May 2, 2000 by presidential order, the effect on surveying applications of GPS was negligible.

Selective Availability (SA) means that the GPS signals were transmitted from the satellites with intentional clock errors added. The stability of the atomic clocks onboard was deliberately degraded and thereby the Navigation message was degraded too. While Selective Availability was removed by Precise Positioning Service, P code, users with decryption techniques, civilian users of the Standard Positioning Service C/A code were not able to remove these errors.

Selective availability was on since the first launch of the Block II satellites in 1989. It was turned off briefly to allow coalition forces in the Persian Gulf to use civilian GPS receivers in 1990, but was turned on again immediately.

For the code phase receiver owner limited to pseudorange positioning, selective availability was a problem. A single pseudorange receiver could only achieve positional accuracy of about ± 100 meters 95% of the time. This sort of single point positioning is not a usual surveying application of GPS. But with a second code phase receiver as a base station on a known position and using differential data processing techniques, submeter accuracy was possible even with SA turned on.

The carrier phase receivers used in most surveying applications have always been, for all practical purposes, immune from the effects of SA.

2

Biases and Solutions

THE ERROR BUDGET

A Look at the Biases in the Observation Equations

The management of errors is indispensable for finding the true geometric range ρ from either a pseudorange, or carrier phase observation.

$$p = \rho + c\left(dt - dT\right) + d_{ion} + d_{trop} + \varepsilon_p \text{ (pseudorange)}$$

$$\Phi = \rho + c\left(dt - dT\right) + \lambda N - d_{ion} + d_{trop} + \varepsilon_\Phi \text{ (carrier phase)}$$

Both equations include environmental and physical limitations called *range biases.*

The Biases

Among these biases are certain atmospheric errors; two such errors are the ionospheric effect, d_{ion}, and the tropospheric effect, d_{trop}. The tropospheric effect may be somewhat familiar to EDM users, even if the ionospheric effect is not. Other biases, clock errors symbolized by $(dt - dT)$ and receiver noise, multipath, etc., (combined in the symbols ε_p and ε_Φ) are unique to satellite surveying methods. The objective here is to mathematically separate and quantify each of these biases.

UERE

When each bias is expressed as a range itself, each quantity is known as a *user equivalent range error* or *UERE*. This expression, often used in GPS literature, is a convenient way to clarify the individual contributions of each bias to the overall biased range measurement (Figure 2.1).

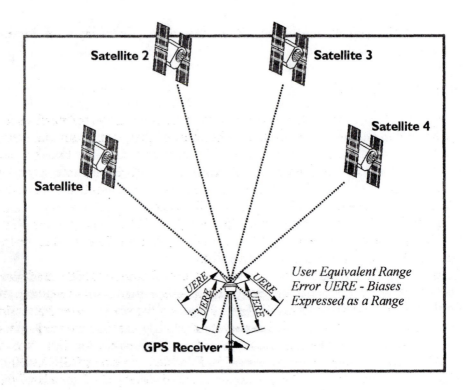

Figure 2.1. User Equivalent Range Error.

The Satellite Clock Bias, *dt*

The largest error can be attributed to the satellite clock bias. It can be quite large if the broadcast clock correction is **not** used by the receiver to bring the time signal acquired from a satellite's onboard clock in line with GPS time.

Relativistic Effects on the Satellite Clock

Albert Einstein's special and general theories of relativity predicted that a clock in orbit around the earth would appear to run faster than a clock on its surface. And they do indeed, due to their greater speed and the weaker gravity around them, the clocks in the GPS satellites do appear to run faster than the clocks in GPS receivers. There are actually two parts to the effect.

Concerning the first part, time dilation is taken into account before the satellites' clocks are sent into orbit. To ensure the clocks will actually achieve the correct fundamental frequency of 10.23 MHz in space, their frequency is set a bit slow before launch to 10.22999999545 MHz.

The second part is attributable to the eccentricity of the orbit of GPS satellites. With an eccentricity of 0.02, this effect can be as much as 45.8 nanoseconds. Fortunately, the offset is eliminated by a calculation in the GPS receiver itself, thereby avoiding a ranging error of about 14 meters.

Both relativistic effects on the satellite clocks can be accurately computed and are removed from the system.

Satellite Clock Drift

Clock drift is another matter. As discussed in Chapter 1, the onboard satellite clocks are independent of one another. The rates of these rubidium and cesium oscillators are more stable if they are not disturbed by frequent tweaking and adjustment is kept to a minimum. While GPS time itself is designed to be kept within one microsecond of UTC, excepting leap seconds, the satellite clocks can be allowed to drift up to a millisecond from GPS time.

There are three kinds of time involved here. The first is UTC per the United States Naval Observatory (USNO). The second is GPS time. The third is the time determined by each independent GPS satellite.

Their relationship is as follows. The Master Control Station (MCS) at Schriever (formerly Falcon) Air Force Base near Colorado Springs, Colorado gathers the GPS satellites' data from monitoring stations around the world. After processing, this information is uploaded back to each satellite to become the broadcast ephemeris, broadcast clock correction, etc. The actual specification for GPS time demands that it be within one microsecond of UTC as determined by USNO, without consideration of leap seconds. Leap seconds are used to keep UTC correlated with the actual rotation of the Earth, but they are ignored in GPS time. In GPS time it is as if no leap seconds have occurred at all in UTC since 24:00:00, January 5, 1980. And in practice, GPS time is much closer than the microsecond specification; it is usually within about 40 nanoseconds of UTC, minus leap seconds.

The system also ensures that the time broadcast by each independent satellite in the constellation is no worse than one millisecond from GPS time. By constantly monitoring the satellites' clock error, dt, the Control Segment gathers data for its daily uploads of the broadcast clock corrections. The primary purpose of these corrections is to reduce an error from around 1 millisecond of satellite clock error to around 30 nanoseconds (that is 30 billionths of a second) of GPS time. You will recall that clock corrections are part of the Navigation message; more specifically, part of subframe 1 of that message.

The Ionospheric Effect, d_{ion}

The relatively unhindered travel of the GPS signal through the virtual vacuum of space changes as it passes through the earth's atmosphere. Through both refraction and diffraction, the atmosphere alters the apparent speed and, to a lesser extent, the direction of the signal.

Group and Phase Delay

The ionosphere is the first layer the signal encounters. It extends from about 50 km to 1000 km above the earth's surface. Traveling through this layer of the atmosphere the most troublesome effects on the GPS signal are known as the *group delay* and the *phase delay*. They both alter the measured range. The magnitude of these delays is determined by the density and stratification of the ionosphere at the moment the signal passes through it.

TEC

The density of the ionosphere changes with the number and dispersion of free electrons released when gas molecules are ionized by the sun's ultraviolet radiation. This density is often described as *total electron content* or *TEC*, a measure of the number of free electrons in a column through the ionosphere with a cross-sectional area of 1 square meter.

Ionosphere and the Sun

During the daylight hours in the midlatitudes the ionospheric delay may be as much as five times greater than it is at night. In fact, the delay is usually least between midnight and early morning and most at about local noon. It is also nearly four times greater in November, when the earth is nearing its *perihelion*, its closest approach to the sun, than it is in July near the earth's *aphelion*, its farthest point from the sun. The effects of the ionosphere on the GPS signal usually reach their peak in March, about the time of the vernal equinox.

Ionospheric Gradients

The ionosphere is not homogeneous. It changes from layer to layer within a particular area. Its behavior in one region of the earth is liable to be unlike its behavior in another. For example, ionospheric disturbances can be particularly harsh in the polar regions. But the highest TEC values and the widest variations in the horizontal gradients occur in the band of about 60° of *geomagnetic latitude*. That band lies 30° north and 30° south of the earth's magnetic equator.

The disturbances in the ionosphere in the equatorial region have a severity not seen in the midlatitudes. In that area, refraction and diffraction peak from sunset to midnight. Their effect may become severe enough to cause receivers to lose their lock on the GPS signal.

Satellite Elevation and Ionospheric Effect

The severity of the ionospheric effect varies with the amount of time the GPS signal spends traveling through the layer. A signal originating from a satellite near the observer's horizon must pass through a larger amount of the ionosphere to reach the receiver than does a signal from a satellite near the observer's zenith. The longer the signal is in the ionosphere, the greater the ionosphere's effect, and the greater the impact of horizontal gradients within the layer.

The Magnitude of the Ionospheric Delay

The error introduced by the ionosphere can be very small; it may be as large as 150 meters when the satellite is near the observer's horizon, the vernal equinox is near, and sunspot activity is at its maximum. It is usually not that severe. It varies with magnetic activity, location, time of day, and even the direction of observation. However, the ionospheric effect can contribute the second largest UERE.

The Ionosphere Affects Codes and the Carrier Differently

Fortunately, the ionosphere has a property that can be used to minimize its effect on GPS signals. It is *dispersive*—that means that the apparent time delay contributed by the ionosphere depends on the frequency of the signal. One result of this dispersive property is that during the signal's trip through the ionosphere, the codes, the modulations on the carrier wave, are affected differently than the carrier wave itself.

The ionospheric delay can be divided into two distinct categories: *phase delay* and *group delay*. All the modulations on the carrier wave, the P code, the C/A code, and the Navigation message, appear to be slowed. They are affected by the group delay. But the carrier wave itself appears to speed up in the ionosphere. It is affected by the phase delay.

It may seem odd to call an increase in speed a delay, but, governed by the same properties of electron content as the group delay, phase delay just increases negatively. Please note that the algebraic sign of d_{ion} is negative in the carrier phase equation and positive in the pseudorange equation.

Different Frequencies Are Affected Differently

Another consequence of the *dispersive* nature of the ionosphere is that the apparent time delay for a higher frequency carrier wave is less than it is for a lower frequency wave. That means that L1, 1575.42 MHz, is not affected as much as L2, 1227.60 MHz. This fact provides one of the greatest advantages of a dual-frequency receiver over the single-frequency receivers. By tracking both carriers, a dual-frequency receiver has the facility of modeling and removing not all, but a significant portion of the ionospheric bias.

The frequency dependence of the ionospheric effect is described by the following expression (Klobuchar, 1983 in Brunner and Welch, 1993):

$$v = \frac{40.3}{cf^2} \cdot \text{TEC}$$

Where　v　=　the ionospheric delay
c　=　the speed of light in meters per second
f　=　the frequency of the signal in Hz
TEC　=　the quantity of free electrons per cubic meter.

As the formula illustrates, the time delay is inversely proportional to the square of the frequency. In other words, the higher the frequency, the less the delay. Hence the dual-frequency receiver's capability to discriminate the effect on L1 from that on L2. Such a dual-frequency model can be used to reduce the ionospheric bias. Still, it is far from perfect and cannot ensure the effect's elimination.

Broadcast Correction

As mentioned in Chapter 1, an ionospheric correction is also available to the single frequency receiver in subframe 4 of the Navigation message. However, this broadcast correction should not be expected to remove more than about three-quarters of the ionospheric effect.

The Receiver Clock Bias, *dT*

The third largest error can be caused by the receiver clock; that is, its oscillator. Both a receiver's measurement of phase differences and its generation of replica codes depend on the reliability of this internal frequency standard.

Typical Receiver Clocks

GPS receivers are usually equipped with quartz crystal clocks which are relatively inexpensive and compact. They have low power requirements and long life spans. For these types of clocks, the frequency is generated by the piezoelectric effect in an oven-controlled quartz crystal disk, a device sometimes symbolized by *OCXO*. Their reliability ranges from a minimum of about 1 part in 10^8 to a maximum of about 1 part in 10^{10}. The latter is a magnitude about equal to a quarter of a second over a human lifetime. Even at that, quartz clocks are not as stable as the atomic standards in the GPS satellites and are more sensitive to temperature changes, shock, and vibration. Some receiver designs augment their frequency standards by also having the capability to accept external timing from cesium or rubidium oscillators.

The Orbital Bias

Orbital bias has the potential to be the fourth largest error, but this ephemeris-based bias is not symbolized in the observation equations at all. It is addressed in the broadcast ephemeris.

Forces Acting on the Satellites

The orbital motion of GPS satellites is not only a result of the earth's gravitational attraction, there are several other forces that act on the satellite. The primary disturbing forces are the nonspherical nature of the earth's gravity, the attractions of the sun and the moon, and solar radiation pressure. The best model of these forces is the actual motion of the satellites themselves, and the government facilities distributed around the world, known collectively as the *Control Segment,* track them for that reason, among others.

Tracking Facilities

The Master Control Station *MCS* is located at Schriever Air Force Base in Colorado Springs, Colorado. The station is manned by the 2nd Space Operations Squadron. It computes updates for the Navigation message, generally, and the broadcast ephemeris, in particular, based on about one week of tracking information collected from five monitoring stations around the world.

The system requires maintenance. Orbital and clock adjustments and other data uploads are necessary to keep the constellation from degrading. For example, GPS satellites are stabilized by *wheels,* a series of gyroscopic devices. Left alone, these devices would continually accelerate until they were useless. Fortunately, the excess momentum is periodically dumped on instructions from the Control Segment.

Every GPS satellite is tracked by at least one monitoring station at all times and the orbital tracking data gathered by monitoring stations are then passed on to the Master Control Station. There, new ephemerides are computed. This tabulation of the anticipated locations of the satellites with respect to time is then transferred to four uploading stations, where it is transmitted back to the satellites themselves.

There are some difficulties in the process of updating the satellite's orbital information. The broadcast ephemeris must be a least-squares prediction based on the satellite's past behavior.

The Tropospheric Effect, d_{trop}

The fifth largest UERE can be attributed to the effect of the troposphere.

Troposphere

The troposphere is that part of the atmosphere closest to the earth. In fact, it extends from the surface to about 9 km over the poles and 16 km over the equator, but in this work it will be combined with the tropopause and the stratosphere, as it is in much of GPS literature. Therefore, the following discussion of the tropospheric effect will include the layers of the earth's atmosphere up to about 50 km above the surface.

Tropospheric effect is independent of frequency. The tropospheric delay appears to add a slight distance to the range the receiver measures between itself and the satellite. The troposphere and the ionosphere are by no means alike in their effect on the satellite's signal. The troposphere is *nondispersive* for frequencies below 30 GHz. In other words, the refraction of a GPS satellite's signal is not related to its frequency in the troposphere.

The troposphere is part of the *electrically neutral* layer of the earth's atmosphere, meaning it is neither ionized nor dispersive. Therefore, both L1 and L2 are equally refracted. Like the ionosphere, the density of the troposphere also governs the severity of its effect on the GPS signal. For example, when a satellite is close to the horizon, the delay of the signal caused by the troposphere is maximized. The tropospheric delay of the signal from a satellite at zenith, directly above the receiver, is minimized.

Satellite elevation and tropospheric effect. The situation is analogous to atmospheric refraction in astronomic observations; the effect increases as the energy passes through more of the atmosphere. The difference in GPS is that it is the delay, not the angular deviation, caused by the changing density of the atmosphere that is of primary interest. The GPS signal that travels the shortest path through the troposphere will be the least delayed by it. So, even though the delay at an elevation angle of 90° will only be about 2.4 meters, it can increase to about 9.3 meters at 75° and up to 20 meters at 10°.

Modeling. Modeling the troposphere is one technique used to reduce the bias in GPS data processing, and it can be up to 95 % effective. However, the residual 5 % can be quite difficult to remove. For example, surface measurements of temperature and humidity are not strong indicators of conditions on the path between the receiver and the satellite. But instruments that can provide some idea of the conditions along the line between the satellite and the receiver are somewhat more helpful in modeling the tropospheric effect.

The dry and wet components of refraction. Refraction in the troposphere has a dry component and a wet component. The dry component is closely correlated to the atmospheric pressure and can be more easily estimated than the wet component. It is fortunate that the dry component contributes the larger portion of range error in the troposphere since the high cost of water vapor radiometers and radiosondes generally restricts their use to only the most high-precision GPS work.

Receiver spacing and the atmospheric biases. There are other practical consequences of the atmospheric biases. For example, the character of the atmosphere is never homogeneous and the importance of atmospheric modeling increases as the distance between GPS receivers grows. Consider a signal traveling from one satellite to two receivers that are close together. That signal would be subjected to very similar atmospheric effects, and, therefore, atmospheric bias modeling would be less important to the accuracy of the measurement of the relative distance between them. But a signal traveling from the same satellite to two receivers that are far apart may pass through levels of atmosphere quite different from one another. In that case, atmospheric bias modeling would be more important.

Multipath

The range delay known as *multipath,* is symbolized by ε_p and ε_Φ. It differs from both the apparent slowing of the signal through the ionosphere and troposphere and the discrepancies caused by clock offsets. The range delay in multipath is the result of the reflection of the GPS signal.

Multipath occurs when part of the signal from the satellite reaches the receiver after reflecting from the ground, a building, or another object. These reflected signals can interfere with the signal that reaches the receiver directly from the satellite.

Limiting the Effect of Multipath

The high frequency of the GPS codes tends to limit the field over which multipath can contaminate pseudorange observations. Once a receiver has achieved lock; that is, its replica code is correlated with the incoming signal from the satellite, signals outside the expected chip length can be rejected.

There are other factors that distinguish reflected multipath signals from direct signals. For example, reflected signals at the frequencies used for L1 and L2 tend to be more diffuse than the directly received signals. Another difference involves the circular polarization of the GPS signal. The polarization is actually reversed when the signal is reflected, similar to the spin of a banked billiard ball. These characteristics allow some multipath signals to be identified and rejected at the receiver's antenna.

Antenna Design and Multipath

GPS antenna design can play a role in minimizing the effect of multipath. Ground planes, usually a metal sheet, are used with many antennas to reduce multipath interference by eliminating signals from low elevation angles.

Choke ring antennas, based on a design first introduced by the Jet Propulsion Laboratory (JPL), can reduce antenna gain at low elevations. This design contains a series of concentric circular troughs that are a bit more than a quarter of a wavelength deep. When a GPS signal's wavefront arrives at the edge of an antenna's ground plane from below, it can induce a surface wave on the top of the plane that travels horizontally. A choke ring antenna can prevent the formation of these surface waves.

But neither ground planes nor choke rings mitigate the effect of reflected signals from above the antenna very effectively. There are signal processing techniques that can reduce multipath, but when the reflected signal originates less than a few meters from the antenna, this approach is not as effective.

A widely used strategy is the 15° cutoff or *mask angle*. This technique calls for tracking satellites only after they are more than 15° above the receiver's horizon. Careful attention in placing the antenna away from reflective surfaces, such as nearby buildings or vehicles, is another way of minimizing the occurrence of multipath.

DIFFERENCING

Classifications of Positioning Solutions

Kinematic GPS

Speaking broadly, concerning carrier phase observations there are two applications of GPS in surveying and geodesy, *kinematic* and *static* positioning. Kinematic applications imply movement, one or more GPS receivers actually in motion during their observations. A moving GPS receiver on land, sea, or air, is characteristic of kinematic GPS. Other characteristics of the application include results in real time and little redundancy. Hydrography, aerial mapping, gravimetric and more and more land surveying projects are done using kinematic GPS. The kinematic GPS surveying used in land surveying today was originated by Dr. Benjamin Remondi in the 1980s.

Static GPS

Static applications use observations from receivers that are stationary for the duration of their measurement. Most static applications can afford higher redundancy and a bit more accuracy than kinematic GPS. The majority of GPS surveying control and geodetic work still relies on static applications.

Hybrid Techniques in GPS

There have been recent hybrids between kinematic and static applications. *Pseudokinematic* and *semikinematic* rely on short static observations and, in the first instance, revisits to positions once occupied; in the second, the alternate movement of multiple receivers. Another hybrid is called *rapid static*. The receivers used in rapid static are stationary during their observations, but their observations are very short, and they rely on both code and carrier observations on both L1 and L2.

Relative and Point Positioning

Kinematic and static applications include two classifications *point, or autonomous positioning* and *relative positioning*. In other words, there are four possible combinations. They are: *static point positioning, static relative positioning, kinematic point positioning,* and *kinematic relative positioning.*

The term *differential* GPS, or *DGPS*, has come into common usage as well. Use of this acronym usually indicates a method of relative positioning where coded pseudorange measurements are used rather than carrier phase.

The Navigation Solution

Another type of point positioning is known as *absolute positioning, single-point positioning,* or *the Navigation solution.* It is characterized by a single receiver measuring its range to a minimum of four satellites simultaneously (Figure 2.2).

The Navigation solution is in one sense the fulfillment of the original idea behind GPS. It relies on a coded pseudorange measurement and can be used for virtually instantaneous positioning.

In this method the positions of the satellites are available from the data in their broadcast ephemerides. The satellite clock offset and the ionospheric correction are also available from the Navigation messages of all four satellites.

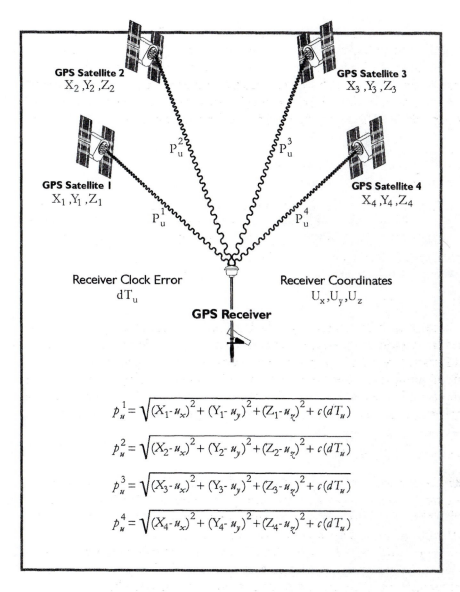

Figure 2.2. Navigation Solution.

Four Unknowns

Even if all these data are presumed to contain no errors, which they surely do, four unknowns remain: the position of the receiver in three Cartesian coordinates, u_x, u_y, and u_z, and the receiver's clock error dT_u. Three pseudoranges provide enough data to solve for u_x, u_y, and u_z. And the fourth pseudorange provides the information for the solution of the receiver's clock offset.

The ability to achieve so much redundancy in the measurement of the dT, the receiver's clock error, is one reason the moderate stability of quartz crystal clock technology is entirely adequate as a receiver oscillator.

Four Satellites and Four Equations

A unique solution is found here because the number of unknowns is not greater than the number of observations. The receiver tracks four satellites simultaneously; therefore, these four equations can be solved simultaneously for every *epoch* of the observation. An epoch in GPS is a very short period of observation time, and is generally just a small part of a longer measurement. However, theoretically there is enough information in any single epoch of the Navigation solution to solve these equations. In fact, the trajectory of a receiver in a moving vehicle could be determined by this method. With four satellites available, resolution of a receiver's position and velocity are both available through the simultaneous solution of these four equations. These facts are the foundation of the kinematic application of GPS.

Relative Positioning

Correlation

The availability of two or more GPS receivers makes relative positioning possible. Relative positioning can attain higher accuracy than point positioning because of the extensive correlation between observations taken to the same satellites at the same time from separate stations. Because the distance between such stations on the earth are short compared with the 20,000-km altitude of the GPS satellites, two receivers operating simultaneously, collecting signals from the same satellites will record very similar errors.

Baselines

The vectors between such pairs of receivers are known as *baselines*. The simultaneity of observation, the resulting error correlation, and the carrier phase observable, combine to yield typical baseline measurement accuracies of ± (1 cm + 2 ppm). It is possible for GPS measurements of baselines to be as accurate as 1 ppm or even 0.1 ppm. For example, a nine mile baseline correctly measured by GPS within ± 0.05 ft (1 ppm) to ± 0.005 ft (0.1 ppm).

Networks

Network and *multireceiver positioning* are obvious extensions of relative positioning. Both the creation of a closed network of points by combining individually observed baselines and the operation of three or more receivers simultaneously have advantages. For example, the baselines have redundant measurements and similar, if not identical, range errors. The processing methods in such an arrangement can nearly eliminate many of the biases introduced by imperfect clocks and the atmosphere. These processing strategies are based on computing the differences between simultaneous GPS carrier phase observations.

Differencing

In GPS the word *differencing* has come to represent several types of simultaneous baseline solutions of combined measurements. The most frequently used are known as the *single difference, double difference,* and *triple difference.*

Single Difference

One of the foundations of differencing is the idea of the baseline as it is used in GPS. For example, a single difference, also known as a *between-receivers difference,* can refer to the difference in the simultaneous carrier phase measurements from one GPS satellite as measured by two different receivers (Figure 2.3).

Elimination of the Satellite Clock Errors

In this single-difference solution the satellite clock error is eliminated. The difference between dt at the first receiver and dt at the second receiver, Δdt, is zero. Since the two receivers are both observing the same satellite at the same time, the satellite clock bias is canceled. The atmospheric biases and the orbital errors recorded by the two receivers in this solution are nearly identical, so they too can also be virtually eliminated.

Other Errors Remain

Unfortunately, there are still two factors in the carrier beat phase observable that are not eliminated by single differencing. The difference between the integer cycle ambiguities at each receiver, ΔN, and the difference between the receiver clock errors, ΔdT, remain.

Double Difference

Elimination of the Receiver Clock Errors

There is a GPS solution that will eliminate the receiver clock errors. It involves the addition of what might be called another kind of single difference, also known as a *between-satellites difference.* This term refers to the difference in the measurement of signals from two GPS satellites, as measured simultaneously at a single receiver (Figure 2.4).

The data available from the between-satellites difference allow the elimination of the receiver clock error. In this situation, there can be no difference in the clock since only one is involved. And the atmospheric effects on the two satellite signals are again nearly identical as they come into the lone receiver, so the effects of the ionospheric and tropospheric delays are virtually eliminated as well.

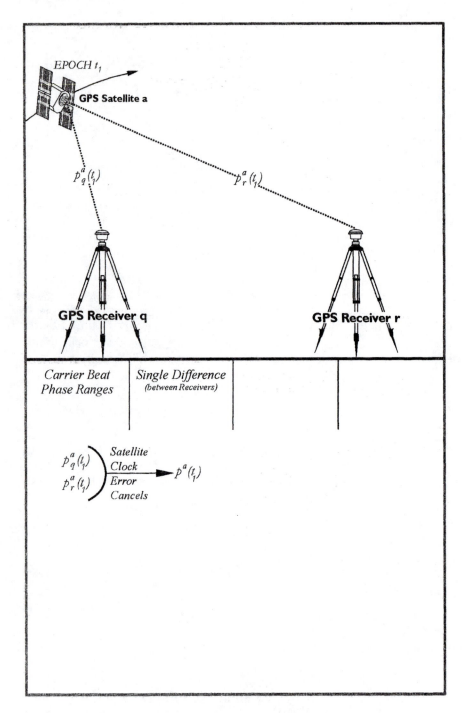

Figure 2.3. Carrier Beat Phase Measurements from One GPS Satellite, as Measured by Two Different Receivers.

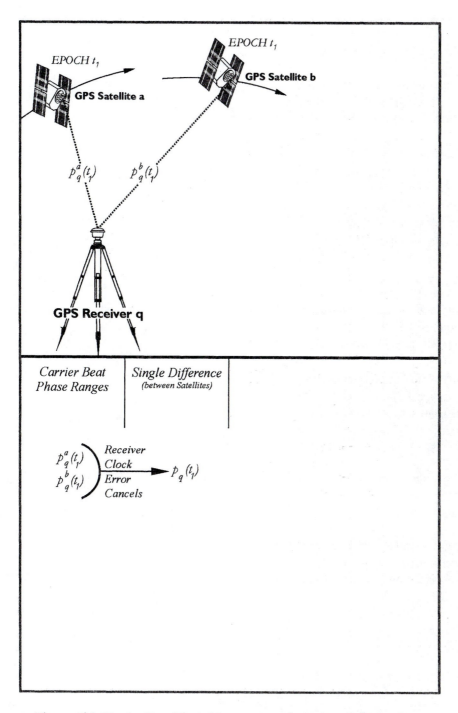

Figure 2.4. Carrier Beat Phase Measurement from Two GPS Satellites, as Measured at a Single Receiver.

No Clock Errors at All

By using both the between-receivers difference and the between-satellites difference, a double difference is created. This combination is virtually free of receiver clock errors. Therefore, for all practical purposes, the double difference does not contain receiver clock errors nor satellite clock errors (Figure 2.5).

Still there is one stubborn factor in the carrier beat phase observable that is not eliminated. The integer cycle, or phase ambiguity, N.

Triple Difference

A third kind of differencing is created by combining two double differences. Each of the double differences involves two satellites and two receivers. The difference next derived is between two epochs. The triple difference is also known as the *receiver-satellite-time triple difference* (Figure 2.6), the difference of two double differences of two different epochs.

Integer Cycle Ambiguity

In the triple difference two receivers observe the same two satellites during two consecutive epochs. This solution can be used to quantify the integer cycle ambiguity, N, because if all is as it should be, N is constant over the two observed epochs. Therefore, the triple difference makes the detection and elimination of cycle slips relatively easy.

Actually, a triple difference is not sufficiently accurate for short baselines. It is used to find the integer cycle ambiguity. Once the cycle ambiguity is determined it can be used with the double difference solution to calculate the actual carrier phase measurement.

Cycle slips. A *cycle slip* is a discontinuity in a receiver's continuous phase lock on a satellite's signal. The coded pseudorange measurement is immune from this difficulty, but the carrier beat phase is not.

Components of the carrier phase observable. From the moment a receiver locks onto a particular satellite, there are actually three components to the total carrier phase observable.

$$\phi = \alpha + \beta + N$$

where: ϕ = total phase
   α = fractional initial phase
   β = observed cycle count
   N = cycle count at lock on

Fractional initial phase. First is the *fractional initial phase,* which occurs at the receiver at the first instant of the lock-on. The receiver starts tracking the incoming phase from the satellite. It cannot yet know how to achieve a perfect synchro-

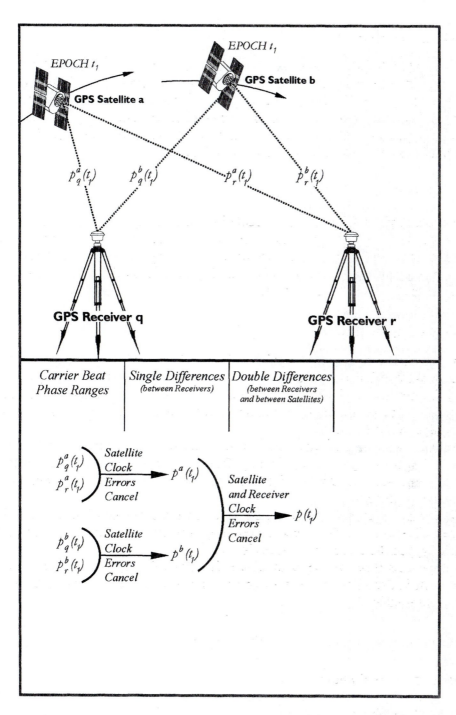

Figure 2.5. Carrier Beat Phase Measurement from Two GPS Satellites, as Measured at Two Receivers.

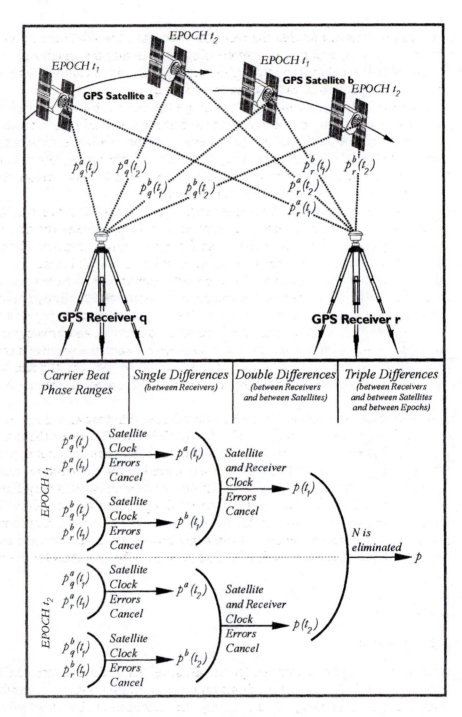

Figure 2.6. The Receiver-Satellite-Time Triple Difference.

nization. Lacking this knowledge the receiver grabs onto the satellite's signal at some fractional part of a phase. It is interesting to note that this fractional part does not change for the duration of the observation and so is called the fractional initial phase. It is symbolized in the equation above by α.

The integer number of cycles. Second is the integer number of full cycles of phase that occur from the moment of the lock to the end of the observation. It is symbolized by β, the observed cycle count. This element is the receiver's consecutive counting of the change in full phase cycles, 1, 2, 3, 4, ..., between the receiver and the satellite as the satellite flies over. Of the three terms, β is only number that changes—that is, if the observation proceeds correctly.

The cycle ambiguity. Third is the integer cycle ambiguity N. It represents the number of full phase cycles between the receiver and the satellite at the first instant of the receiver's lock-on. It is labeled as the cycle count at lock-on. N does not change from the moment of the lock onward, unless that lock is lost.

In other words, the total carrier phase observable consists of two values that do not change during the observation, the fractional phase α, and the integer cycle ambiguity N. Only the observed cycle count β, changes, unless there is a cycle slip.

Lost lock. When lock is lost a cycle slip occurs. A power loss, an obstruction, a very low signal-to-noise ratio, or any other event that breaks the receiver's continuous reception of the satellite's signal causes a cycle slip. That is, the receiver loses its place in its count of the integer number of cycles β and, as a result, N is completely lost.

Finding and fixing cycle slips. There are several methods that may be used to regain a lost integer phase value, N. The triple difference is one of the better alternatives in this regard; as stated earlier, the triple difference does not depend on the initial integer ambiguity, because it is a constant in time. Therefore, when a large residual does appear in its component double differences it is very likely caused by a cycle slip. Even better, the obstructed signal can be singled-out by isolating all available satellite pairs until the problem is found. This utility in fixing cycle slips is the primary appeal of the triple difference. It can be used as a preprocessing step to weed out cycle slips and provide a first position for the receivers.

Summary

Typical Techniques

Relative positioning by carrier beat phase measurement is the primary vehicle for high-accuracy GPS surveying. Simultaneous static observations, double differencing in postprocessing, and the subsequent construction of networks from GPS baselines are the hallmarks of geodetic and control work in the field. The strengths of these methods generally outweigh their weaknesses, particularly where there can be an unobstructed sky and relatively short baselines and where the length of observation sessions is not severely restricted.

Pseudorange and Carrier Phase

However, conditions are not always so ideal. Where obstructions threaten to produce cycle slips, coded pseudorange measurements may offer an important advantage over carrier beat phase. Pseudorange measurements also may be preferred where accuracy requirements are low and production demands are high.

Kinematic GPS

There can hardly be a question that kinematic GPS is the most productive of the several alternative methods, under the right circumstances. However, the necessity of maintaining lock on four or more satellites as the receiver is moved, currently limits its application to very open areas.

Other Techniques

Rapid-static, pseudokinematic, and other hybrid methods are attempts to take advantage of some of the best aspects of static positioning, such as high accuracy and predictable production, while improving on its drawbacks. It is almost always desirable to increase production by employing shortened observation sessions, provided it can be done while maintaining the required accuracy.

Differencing

Differencing is an ingenious approach to minimizing the effect of errors in carrier beat phase ranging. It is a technique that largely overcomes the impossibility of perfect time synchronization. Double differencing is the most widely used formulation. Double differencing still contains the initial integer ambiguities, of course. And the estimates of the ambiguities generated by the initial processing are usually not integers; in other words, some orbital errors, atmospheric errors, etc., remain. But with the knowledge that the ambiguities ought to be integers, during subsequent processing it is possible to force estimates for the ambiguities that are in fact integers. When the integers are so fixed, the results are known as a *fixed solution*, rather than a *float solution*. It is the double-differenced carrier-phase-based fixed solution that makes the very high accuracy possible with GPS.

Biases

However, in this discussion of errors it is important to remember that multipath, cycle slips, incorrect instrument heights, and a score of other errors whose effects can be minimized or eliminated by good practice are simply not within the purview of differencing at all. The unavoidable biases that can be managed by differencing—including clock, atmospheric, and orbital errors—can have their effects drastically reduced by the proper selection of baselines, the optimal length of the observation sessions, and several other considerations included in the design of a GPS survey. But such decisions require an understanding of the sources of these biases and the conditions that govern their magnitudes. The adage of, "garbage in, garbage out," is as true of GPS as any other surveying procedure. The management of errors cannot be relegated to mathematics alone.

EXERCISES

1. The earth's atmosphere affects the signals from GPS satellites as they pass through. Which of these statements about the phenomenon is true?

   (a) The GPS signals are refracted by the ionosphere. Considering pseudoranges, the apparent length traveled by a GPS signal seems to be too short and in the case of the carrier wave it seems to be too long.

   (b) The ionosphere is dispersive. Therefore, despite the number of charged particles the GPS signal encounters on its way from the satellite to the receiver, the time delay is directly proportional to the square of the transmission frequency. Lower frequencies are less affected by the ionosphere.

   (c) The ionosphere is a region of the atmosphere where free electrons do not affect radio wave propagation, due mostly to the effects of solar ultraviolet and x-ray radiation.

   (d) The TEC depends on the user's location, observing direction, the magnetic activity, the sunspot cycle, the season, and the time of day, among other things. In the midlatitudes the ionospheric effect is usually least between midnight and early morning and most at local noon.

2. The clocks in GPS satellites are rubidium and cesium oscillators, whereas quartz crystal oscillators provide the frequency standard in most GPS receivers. Concerning the relationship between these standards, which of the following statements is <u>not</u> true?

   (a) Quartz clocks are not as stable as the atomic standards in the GPS satellites and are more sensitive to temperature changes, shock, and vibration.

   (b) Both GPS receivers and satellites rely on their oscillators to provide a stable reference so that other frequencies of the system can be generated from or compared with them.

   (c) The foundation of the oscillators in GPS receivers and satellites is the piezoelectric effect.

   (d) The oscillators in the GPS satellites are also known as atomic clocks.

3. What is an advantage available using a dual-frequency GPS receiver that is <u>not</u> available using a single-frequency GPS receiver?

   (a) A single-frequency GPS receiver cannot collect enough data to perform single, double, or triple difference solutions.

   (b) A dual-frequency receiver affords an opportunity to track the P code, but a single-frequency receiver cannot.

   (c) A dual-frequency receiver has access to the Navigation code, but a single-frequency receiver does not.

   (d) Over long baselines, a dual-frequency receiver has the facility of modeling and virtually removing the ionospheric bias, whereas a single-frequency receiver cannot.

4. Which of the following statements is <u>not</u> correct concerning refraction of the GPS signal in the troposphere?

   (a) L1 and L2 carrier waves are refracted equally.

   (b) When a GPS satellite is near the horizon its signal is most affected by the atmosphere.

   (c) The density of the troposphere governs the severity of its effect on a GPS signal.

   (d) The wet component of refraction in the troposphere contributes the larger portion of the range error.

5. In a between-receivers single difference across a short baseline, which of the following problems are <u>not</u> virtually eliminated?

   (a) satellite clock errors
   (b) atmospheric bias
   (c) integer cycle ambiguities
   (d) orbital errors

6. In a double-difference across a short baseline, which of the following problems are <u>not</u> virtually eliminated?

   (a) atmospheric bias
   (b) satellite clock errors
   (c) receiver clock errors
   (d) integer cycle ambiguity

7. Which of the following statements would <u>not</u> correctly complete the sentence beginning, "If cycle slips occurred in the observations . . . "?

   (a) ". . . they could have been caused by intermittent power to the receiver."

(b) ". . . they might be detected by triple differencing."
(c) ". . . they are equally troublesome in pseudorange and carrier phase measurements."
(d) ". . . it is best if they are repaired before double differencing is done."

8. What is the correct value for the maximum time interval allowed between GPS time and a particular GPS satellite's onboard clock?

(a) one nanosecond
(b) one millisecond
(c) one microsecond
(d) one femtosecond

9. Which of the following biases are not mitigated by using relative positioning GPS or DGPS?

(a) the ionospheric effect
(b) the tropospheric effect
(c) multipath
(d) satellite clock errors

10. Which of the following does the Control Segment not compile for upload to the GPS satellites?

(a) almanac information
(b) ephemeris information
(c) satellite clock corrections
(d) P and C/A codes

ANSWERS AND EXPLANATIONS

1. Answer is (d)

Explanation: The modulations on radio carrier waves encountering the free electrons in the ionosphere are retarded, but the carrier wave itself is advanced. The slowing is known as the ionospheric delay. And in the case of pseudoranges the apparent length of the path of the signal is stretched; it seems to be too long. However, when considering the carrier wave the path seems to be too short. It is interesting that the absolute value of these apparent changes is almost exactly equivalent.

The ionosphere is dispersive. Refraction in the ionosphere is a function of a wave's frequency. The higher the frequency the less the refraction and the shorter the delay in the case of modulations on the carrier wave. The ionospheric effect is proportional to the inverse of the frequency squared.

The magnitude of the ionospheric effect increases with electron density. Also known as the total electron content or TEC, the electron density varies in both time and space. However, in the midlatitudes the minimum most often occurs from midnight to early morning and the maximum near local noon.

2. Answer is (c)

Explanation: Rubidium and cesium atomic clocks are aboard the current constellation of GPS satellites. Their operation is based on the resonant transition frequency of the Rb-87 and CS-133 atoms, respectively. On the other hand, the quartz crystal oscillators in GPS receivers utilize the piezoelectric effect. Both GPS receivers and satellites rely on their oscillators to provide a stable reference so that other frequencies of the system can be generated from or compared with them.

3. Answer is (d)

Explanation: Sufficient data to calculate single, double, and triple differences can be available from both single- and dual-frequency receivers. The permission to track the P code is not restricted by single- or dual-frequency capability, but national security. GPS receivers have access to the Navigation code whether they track L1, L2, or both.

Using single-frequency receivers ionospheric corrections can be computed from an ionospheric model. GPS satellites transmit coefficients for ionospheric corrections that most receiver's software use. The models assume a standard distribution of the total electron count; still, with this strategy about a quarter of the ionosphere's actual variance will be missed. But when receivers are close together, say 30 kilometers or less, for all practical purposes the ionospheric delay and carrier phase advance is the same for both receivers. Therefore, phase difference observations over short baselines are little affected by ionospheric bias with single frequency receivers.

However, that is not the case over long baselines of, say, 100 kilometers. Dual-frequency receivers can utilize the dispersive nature of the ionosphere to overcome the effects. The resulting time delay is inversely proportional to the square of the transmission frequency. That means that L1, 1575.42 MHz is not affected as much as L2, 1227.60 MHz. By tracking both carriers, a dual-frequency receiver has the facility of modeling and virtually removing much of the ionospheric bias.

4. Answer is (d)

Explanation: The tropospheric effect is nondispersive for frequencies under 30 GHz. Therefore, it affects both L1 and L2 equally. Refraction in the troposphere has a dry component and a wet component. The dry component is related to the atmospheric pressure and contributes about 90 % of the effect. It is more easily modeled than the wet component. The GPS signal that travels the shortest path through the atmosphere will be the least affected by it. Therefore, the tropospheric delay is least at the zenith and most near the horizon. GPS receivers at the ends of short baselines collect signals that pass through substantially the same atmosphere, and the tropospheric delay may not be troublesome. However, the atmosphere may be very different at the ends of long baselines.

5. Answer is (c)

Explanation: If two or more receivers collect carrier-phase observations from a constellation of satellites, observations that are subsequently loaded into a computer for processing, the first step is usually computation of single differences for each epoch. A between-receivers single difference is the difference in the simultaneous carrier phase measurements from one GPS satellite as measured by two different receivers during a single epoch. Single differences are virtually free of satellite clock errors. The atmospheric biases and the orbital errors recorded by the two receivers in this solution are nearly identical, so they too can be virtually eliminated. However, processing does not usually end at single differences in surveying applications because the difference between the integer cycle ambiguities at each receiver and the difference between the receiver clock errors, remain in the solution.

6. Answer is (d)

Explanation: The differences of two single differences in the same epoch using two satellites is known as a double difference. The combination is virtually free of receiver clock errors, satellite clock errors, and over short baselines, orbital and atmospheric biases. However, the integer cycle ambiguities remain.

7. Answer is (c)

Explanation: A cycle slip is a discontinuity in a receiver's continuous phase lock on a satellite's signal. When lock is lost, a cycle slip occurs. A power loss, an obstruction, a very low signal-to-noise ratio, or any other event that breaks the receiver's continuous reception of the satellite's signal causes a cycle slip. The coded pseudorange measurement is immune from this difficulty, but carrier phase measurements are not. It is best if cycle slips are removed before the double difference solution. When a large residual appears the component double differences of a triple difference it is very likely caused by a cycle slip. This utility in finding and fixing cycle slips is the primary appeal of the triple difference. It can be used as a preprocessing step to weed out cycle slips and provide a first position for the receivers.

8. Answer is (b)

Explanation: The GPS time system is composed of the Master Control Clock and the GPS satellite clocks and is measured from 24:00:00, January 5, 1980. But, while GPS time itself is kept within one microsecond of UTC, excepting leap seconds, the satellite clocks can be allowed to drift up to a millisecond from GPS time. The rates of these onboard rubidium and cesium oscillators are more stable if they are not disturbed by frequent tweaking and adjustment is kept to a minimum.

9. Answer is (c)

Explanation: Two or more GPS receivers make relative positioning and DGPS possible. These techniques can attain high accuracy of the extensive correlation between observations taken to the same satellites at the same time from separate stations. The distance between such stations on the earth are short compared with the 20,000-km altitude of the GPS satellites; two receivers operating simultaneously, collecting signals from the same satellites and through substantially the same atmosphere will record very similar errors of several categories. However, multipath is so dependent on the geometry of the particular location of a receiver it cannot be lessened in this way. Keeping the antenna from reflective surfaces, the use of a mask angle, the use of a ground plane or choke ring antenna, etc., are methods used to reduce multipath errors.

10. Answer is (d)

Explanation: The Master Control Station *MCS* of the Control Segment is located at Schriever Air Force Base in Colorado Springs, Colorado. The station is manned by the 2nd Space Operations Squadron. Almanac and ephemeris information are uploaded by the Control Segment. The ephemeris is informed by orbital perturbations of the satellites. The locations of the other satellites in the constellation are included in the almanac. The Control Segment also provides each satellite's clock correction for rebroadcast. However, the P and C/A codes are actually generated in the satellites themselves.

3

The Framework

TECHNOLOGICAL FORERUNNERS

Consolidation

In the early 1970s the Department of Defense, DOD, commissioned a study to define its future positioning needs. That study found nearly 120 different types of positioning systems in place, all limited by their special and localized requirements. The study called for consolidation and NAVSTAR GPS (navigation system with timing and ranging, global positioning system) was proposed. Specifications for the new system were developed to build on the strengths and avoid the weaknesses of its forerunners. Here is a brief look at the earlier systems and their technological contributions toward the development of GPS.

Terrestrial Radio Positioning

Radar

Long before the satellite era the developers of *RADAR* (radio detecting and ranging) were working out many of the concepts and terms still used in electronic positioning today. For example, the classification of the radio portion of the electromagnetic spectrum by letters, such as the L-band now used in naming the GPS carriers, was introduced during World War II to maintain military secrecy about the new technology.

Actually, the 23-cm wavelength that was originally used for search radar was given the L designation because it was long compared to the shorter 10-cm wavelengths introduced later. The shorter wavelength was called S-band, the S for short. But the Germans used even shorter wavelengths of 1.5 cm. They were called K-band, for kurtz, meaning short in German. Wavelengths that were neither long nor short were given the letter C, for compromise, and P-band, for previous, was used to refer to the very first meter-length wavelengths used in radar. There is also an X-band radar used in fire-control radars and other applications.

In any case, the concept of measuring distance with electromagnetic signals (ranging in GPS) had one of its earliest practical applications in radar. Since then there have been several incarnations of the idea.

Distance by Timing

Shoran (short range navigation), a method of electronic ranging using pulsed *VHF* (very-high-frequency) signals, was originally designed for bomber navigation, but was later adapted to more benign uses. The system depended on a signal, sent by a mobile transmitter-receiver-indicator unit being returned to it by a fixed transponder. The elapsed time of the round-trip was then converted to distances.

Shoran Surveying

It wasn't long before the method was adapted for use in surveying. Using shoran from 1949 to 1957 Canadian geodesists were able to achieve precisions as high as 1:56,000 on lines of several hundred kilometers. Shoran was used in hydrographic surveys in 1945 by the Coast and Geodetic Survey. In 1951 shoran was used to locate islands off Alaska in the Bering Sea that were beyond positioning by visual means. Also in the early '50s the United States Air Force created a shoran-measured trilateration net between Florida and Puerto Rico that was continued on to Trinidad and South America.

Shoran's success led to the development of *Hiran* (high-precision shoran). Its pulsed signal was more focused, its amplitude more precise, and its phase measurements more accurate.

Hiran Surveying

Hiran, also applied to geodesy, was used to make the first connection between Africa, Crete, and Rhodes in 1943. But its most spectacular applications were the arcs of triangulation joining the North American Datum 1927 with the European Datum 1950 in the early '50s. By knitting together continental datums, hiran might be considered to be the first practical step toward positioning on a truly global scale.

Sputnik

These and other radio navigation systems proved that ranges derived from accurate timing of electromagnetic radiation were viable. But useful as they were in geodesy and air navigation, they only whet the appetite for a higher platform. In 1957, the development of Sputnik, the first earth-orbiting satellite, made that possible.

Some of the benefits of earth-orbiting satellites were immediately apparent. It was clear that the potential coverage was virtually unlimited. But other advantages were less obvious.

Satellite Advantages

One advantage is that satellite technology allowed a more flexible choice of frequencies. The coverage of a terrestrial radio navigation system is limited by the propagation characteristics of electromagnetic radiation near the ground. To achieve long ranges, the basically spherical shape of the earth favors low frequencies that stay close to the surface. One system, *Loran-C* (long-range navigation-C), can be used to determine positions up to 3,000 km from a fixed transmitter, but its frequency must be in the *LF* (low-frequency range) at 100 kHz or so. *Omega*, another hyperbolic radio navigation system, can be used at ranges of 9,000 km, but its 10- to 14-kHz frequency is so low it's actually audible. (The range of human hearing is about 20 Hz to 15 kHz). These low frequencies have drawbacks. They can be profoundly affected by unpredictable ionospheric disturbances. And modeling the reduced propagation velocity of a radio signal over land can be difficult. But earth-orbiting satellites could allow the use of a broader range of frequencies, and signals emanating from space are simply more reliable.

Using satellites one could use high-frequency signals to achieve virtually limitless coverage. But development of the technology for launching transmitters with sophisticated frequency standards into orbit was not accomplished immediately. Therefore, some of the earliest extraterrestrial positioning was done with optical systems.

Optical Systems

Optical tracking of satellites is a logical extension of astronomy. The astronomic determination of a position on the earth's surface from star observations, certainly the oldest method, is actually very similar to extrapolating the position of a satellite from a photograph of it crossing the night sky. In fact, the astronomical coordinates, *right ascension* α and *declination* δ, of such a satellite image are calculated from the background of fixed stars.

Triangulation with Photographs

Photographic images that combine reflective satellites and fixed stars are taken with *ballistic cameras* whose chopping shutters open and close very fast. The technique causes both the satellites, illuminated by sunlight or earth-based beacons, and the fixed stars to appear on the plate as a series of dots. Comparative analysis of photographs provides data to calculate the orbit of the satellite. Photographs of the same satellite made by cameras thousands of kilometers apart can thus be used to determine the camera's positions by triangulation. The accuracy of such networks has been estimated as high as ± 5 meters.

Laser Ranging

Other optical systems are much more accurate. One called *SLR* (satellite laser ranging) is similar to measuring the distance to a satellite using a sophisticated

EDM. A laser aimed from the earth to satellites equipped with retro reflectors yields the range. It is instructive that two current GPS satellites carry onboard corner cube reflectors for exactly this purpose. The GPS space vehicles numbered 36 (PRN 06) and 37 (PRN 07), launched in 1994 and 1993 respectively, can be tracked with SLR, thereby allowing ground stations to separate the effect of errors attributable to satellite clocks from errors in the satellite's ephemerides.

The same technique, called *LLR* (lunar laser ranging) is used to measure distances to the moon using corner cube reflectors left there during manned missions. These techniques can achieve positions of centimeter precision when information is gathered from several stations. However, one drawback is that the observations must be spread over long periods, up to a month, and they, of course, depend on two-way measurement.

Optical Drawbacks

While some optical methods, like SLR, can achieve extraordinary accuracies, they can at the same time be subject to some chronic difficulties. Some methods require skies to be clear simultaneously at widely spaced sites. Even then, local atmospheric turbulence causes the images of the satellites to scintillate. The bulky equipment is expensive and optical refraction is difficult to model. In the case of photographic techniques, emulsions cannot be made completely free from irregularities. Still, optical tracking remains a significant part of the satellite management programs of NASA and other agencies.

Extraterrestrial Radio Positioning

Satellite Tracking

The earliest American extraterrestrial systems were designed to assist in satellite tracking and satellite orbit determination, not geodesy. Some of the methods used did not find their way into the GPS technology at all. Some early systems relied on the reflection of signals, transmissions from ground stations that would either bounce off the satellite or stimulate onboard transponders. But systems that required the user to broadcast a signal from the earth to the satellite were not favorably considered in designing the then-new GPS system. Any requirement that the user reveal his position was not attractive to the military planners responsible for developing GPS. They favored a passive system that allowed the user to simply receive the satellite's signal. So, with their two-way measurements and utilization of several frequencies to resolve the cycle ambiguity, many early extraterrestrial tracking systems were harbingers of the modern EDM technology more than GPS.

Prime Minitrack

But elsewhere there were ranging techniques useful to GPS. NASA's first satellite tracking system, *Prime Minitrack,* relied on phase difference measurements of a single-frequency carrier broadcast by the satellites themselves and received by two separate ground-based antennas. This technique is called *interferometry*. Interferometry is the measurement of the difference between the phases of signals that originate at a common source but travel different paths to the receivers. The combination of such signals, collected by two separate receivers, invariably finds them out of step since one has traveled a longer distance than the other. Analysis of the signal's phase difference can yield very accurate ranges and interferometry has become an indispensable measurement technique in several scientific fields.

VLBI

For example, very long baseline interferometry (VLBI) did not originate in the field of satellite tracking or aircraft navigation but in radio astronomy. The technique was so successful it is still in use today. Radio telescopes, sometimes on different continents, tape-record the microwave signals from quasars, starlike points of light billions of light-years from earth (Figure 3.1).

These recordings are encoded with time tags controlled by hydrogen masers, the most stable of oscillators (clocks). The tapes are then brought together and played back at a central processor. Cross-correlation of the time tags reveals the difference in the instants of wavefront arrivals at the two telescopes. The discovery of the time offset that maximizes the correlation of the signals recorded by the two telescopes yields the distance and direction between them within a few centimeters, over thousands of kilometers.

VLBI's potential for geodetic measurement was realized as early as 1967. But the concept of high-accuracy baseline determination using phase differencing was really proven in the late '70s. A direct line of development leads from the VLBI work of that era by a group from the Massachusetts Institute of Technology to today's most accurate GPS ranging technique, carrier phase measurement. VLBI, along with other extraterrestrial systems like SLR, also provide valuable information on the earth's gravitational field and rotational axis. Without that data, the high accuracy of the modern coordinate systems that are critical to the success of GPS, like the Conventional Terrestrial System (CT) would not be possible. But the foundation for routine satellite-based geodesy actually came even earlier and from a completely different direction. The first prototype satellite of the immediate precursor of the GPS system that was successfully launched reached orbit on June 29, 1961. Its range measurements were based on the Doppler effect, not phase differencing, and the system came to be known as *TRANSIT*, or more formally the *Navy Navigational Satellite System*.

TRANSIT

The Doppler Shift

Satellite technology and the Doppler effect were combined in the first comprehensive earth-orbiting satellite system dedicated to positioning. By tracking Sput-

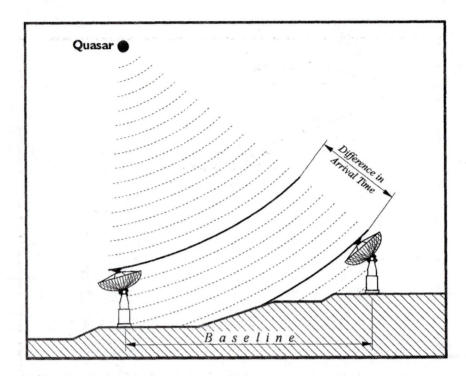

Figure 3.1. Very Long Baseline Interferometry.

nik in 1957, experimenters at Johns Hopkins University's Applied Physics Laboratory found that the Doppler shift of its signal provided enough information to
determine the exact moment of its closest approach to the earth. This discovery
led to the creation of the Navy Navigational Satellite System (NNSS) and the
subsequent launch of six satellites specifically designed to be used for navigation
of military aircraft and ships. This same system, eventually known as TRANSIT,
was classified in 1964, declassified in 1967, and was widely used in civilian surveying for many years until it was switched off on December 31, 1996 (Figure
3.2).

TRANSIT Shortcomings

The TRANSIT system had some nagging drawbacks. For example, its primary
observable was based on the comparison of the nominally constant frequency
generated in the receiver with the Doppler-shifted signal received from one satellite at a time. With a constellation of only six satellites, this strategy sometimes
left the observer waiting up to 90 minutes between satellites and at least two
passes were usually required for acceptable accuracy. With an orbit of only about
1,100 km above the earth, the TRANSIT satellite's orbits were quite low and,
therefore, unusually susceptible to atmospheric drag and gravitational variations,
making the calculation of accurate orbital parameters particularly difficult.

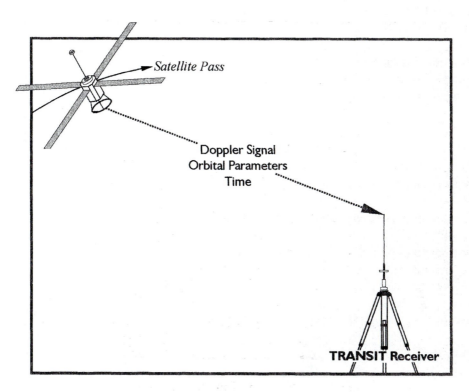

Figure 3.2. TRANSIT Satellite System.

TRANSIT and GPS

Through the decades of the '70s and '80s both the best and the worst aspects of the TRANSIT system were instructive. Some of the most successful strategies of the TRANSIT system have been incorporated into GPS. In both systems the satellites broadcast their own ephemerides to the receivers. Both systems are divided into three segments: the control segment, including the tracking and upload facilities; the space segment, meaning the satellites themselves; and the user segment, everyone with receivers. In both systems the satellites broadcast two frequencies to allow compensation for the ionospheric dispersion. TRANSIT satellites used the frequencies of 400 MHz and 150 MHz, while GPS uses 1575.42 MHz and 1227.60 MHz. The Doppler effect is utilized in both systems. And in both the TRANSIT and GPS systems each satellite and receiver contains its own frequency standards.

Linking Datums

Perhaps the most significant difference between the TRANSIT system and previous extraterrestrial systems was TRANSIT's capability of linking national and international datums with relative ease. Its facility at strengthening geodetic coordinates laid the groundwork for modern geocentric datums.

NAVSTAR GPS

Orbits and Clocks

In 1973, the early GPS experiments were started. From the beginning of GPS, the plan was to include the best features and improve on the shortcomings of all of the previous work in the field. For example, the GPS satellites have been placed in nearly circular orbits over 20,000 km above the earth where the consequences of gravity and atmospheric drag are much less severe. And while the low frequency signals of the TRANSIT system made the ionospheric delay very troublesome, the much higher frequency of GPS signals reduce the effect. The rubidium, cesium, and hydrogen maser time clocks in GPS satellites are a marked improvement over the quartz oscillators used in TRANSIT satellites.

Increased Accuracy

TRANSIT's shortcomings restricted the practical accuracy of the system. Submeter work could only be achieved with long occupation on a station (at least a day) augmented by the use of precise ephemerides for the satellites in postprocessing. GPS provides much more accurate positions in a much shorter time than any of its predecessors, but these improvements are only accomplished by standing on the shoulders of the technologies that have gone before (Table 3.1).

Military Application

The genesis of GPS was military. It grew out of the congressional mandate issued to the Departments of Defense and Transportation to consolidate the myriad of navigation systems. Its application to civilian surveying was not part of the original design. In 1973 the DOD directed the Joint Program Office *(JPO)* in Los Angeles to establish the GPS system. Specifically, JPO was asked to create a system with high accuracy and continuous availability in real time that could provide virtually instantaneous positions to both stationary and moving receivers—all features that the TRANSIT system could not supply.

Secure, Passive, and Global

Worldwide coverage and positioning on a common coordinate grid were required of the new system—a combination that had been difficult, if not impossible, with terrestrial-based systems. It was to be a passive system, which ruled out any transmissions from the users, as had been tried with some previous satellite systems. But the signal was to be secure and resistant to jamming, so codes in the satellite's broadcasts would need to be complex.

Table 3.1. Technologies Preceding GPS

Name of System	Range	Positional Accuracy in Meters	Features
OMEGA	Worldwide	2,200 CEP	Susceptible to VLF propagation anomalies.
LORAN-C	U.S. and selected overseas	180 CEP	Skywave interference and localized coverage.
TRANSIT	Worldwide	Submeter in days	Long waits between satellite passes.
GPS	Worldwide	Centimeter in minutes	24 hour worldwide all weather.

CEP - *Circular Error Probable*

Expense and Frequency Allocation

The DOD also wanted the new system to be free from the sort of ambiguity problems that had plagued OMEGA and other radar systems. And DOD did not want the new system to require large expensive equipment like the optical systems. Finally, frequency allocation was a consideration. The replacement of existing systems would take time, and with so many demands on the available radio spectrum, it was going to be hard to find frequencies for GPS.

Large Capacity Signal

Not only did the specifications for GPS evolve from the experience with earlier positioning systems, so did much of the knowledge needed to satisfy them. Providing 24-hour real-time, high-accuracy navigation for moving vehicles in three dimensions was a tall order. Experience showed part of the answer was a signal that was capable of carrying a very large amount of information efficiently and that required a large bandwidth. So, the GPS signal was given a double-sided 10-MHz bandwidth. But that was still not enough, so the idea of simultaneous observation of several satellites was also incorporated into the GPS system to accommodate the requirement. That decision had far-reaching implications.

The Satellite Constellation

Unlike some of its predecessors, GPS needed to have not one, but at least four satellites above an observer's horizon for adequate positioning. Even more if possible. And the achievement of full-time worldwide GPS coverage would require this condition be satisfied at all times, anywhere on or near the earth.

Spread Spectrum Signal

The specification for the GPS system required all-weather performance and correction for ionospheric propagation delay. TRANSIT had shown that could be accomplished with a dual-frequency transmission from the satellites, but it had also proved that a higher frequency was needed. The GPS signal needed to be secure and resistant to both jamming and multipath. A spread spectrum, meaning spreading the frequency over a band wider than the minimum required for the information it carries, helped on all counts. This wider band also provided ample space for pseudorandom noise encoding, a fairly new development at the time. The PRN codes allowed the GPS receiver to acquire signals from several satellites simultaneously and still distinguish them from one another.

The Perfect System?

The absolutely ideal navigational system, from the military point of view, was described in the Army POS/NAV Master Plan in 1990. It should have worldwide and continuous coverage. The users should be passive. In other words, they should not be required to emit detectable electronic signals to use the system. The ideal system should be capable of being denied to an enemy and it should be able to support an unlimited number of users. It should be resistant to countermeasures and work in real time. It should be applicable to joint and combined operations. There should be no frequency allocation problems. It should be capable of working on common grids or map datums appropriate for all users. The positional accuracy should not be degraded by changes in altitude nor by the time of day or year. Operating personnel should be capable of maintaining the system. It should not be dependent on externally generated signals and it should not have decreasing accuracy over time or the distance traveled. Finally, it should not be dependent on the identification of precise locations to initiate or update the system.

A pretty tall order, and GPS lives up to most, though not all, of the specifications.

GPS in Civilian Surveying

As mentioned earlier, application to civilian surveying was not part of the original concept of GPS. The civilian use of GPS grew up through partnerships between various public, private, and academic organizations. Nevertheless, while the military was still testing its first receivers, GPS was already in use by civilians. Geodetic surveys were actually underway with commercially available receivers early in the 1980s.

Federal Specifications

The Federal Radionavigation Plan of 1984, a biennial document including official policies regarding radionavigation, set the predictable and repeatable accuracy of civil and commercial use of GPS at 100 meters horizontally and 156 meters vertically. This specification meant that the C/A code ranging for Standard Positioning Service could be defined by a horizontal circle with a radius of 100 meters, 95 % of the time. However, that same year civilian users were already achieving results up to six orders of magnitude better than that limit.

And since that time the accuracy available from SPS has increased from ± 100 meters, horizontally, 95 % of the time, to about ± 20 meters to ± 40 meters since Selective Availability was switched off on May 2, 2000 by presidential order.

Interferometry

By using interferometry, the technique that had worked so well with Prime Minitrack and VLBI, civilian users were showing that GPS surveying was capable of extraordinary results. In the summer of 1982 a research group at the Massachusetts Institute of Technology (MIT) tested an early GPS receiver and achieved accuracies of 1 and 2 ppm of the station separation. Over a period of several years extensive testing was conducted around the world that confirmed and improved on these results. In 1984 a GPS network was produced to control the construction of the Stanford Linear Accelerator. This GPS network provided accuracy at the millimeter level. In other words, by using the carrier phase observable instead of code ranging, private firms and researchers were going far beyond the accuracies the U.S. government expected to be available to civilian users of GPS.

The interferometric solutions made possible by computerized processing developed with earlier extraterrestrial systems were applied to GPS by the first commercial users. The combination made the accuracy of GPS its most impressive characteristic, but it hardly solved every problem. For many years, the system was restricted by the shortage of satellites as the constellation slowly grew. The necessity of having four satellites above the horizon restricted the available observation sessions to a few, sometimes inconvenient, windows of time. Another drawback of GPS for the civilian user was the cost and the limited application of both the hardware and the software. GPS was too expensive and too inconvenient for everyday use, but the accuracy of GPS surveying was already extraordinary in the beginning.

Civil Applications of GPS

Today, with a mask angle of 10°, there are some periods when up to 10 satellites are above the horizon. And GPS receivers have grown from only a handful to the huge variety of receivers available today. Some push the envelope to achieve ever-higher accuracy; others offer less sophistication and lower cost. The civilian user's options are broader with GPS than with any previous satellite positioning system—so broad that, as originally planned, GPS will likely replace more of its predecessors in both the military and civilian arenas, as it has replaced the TRANSIT system. In fact, GPS has developed into a system that has more civilian applications and users than military ones. But the extraordinary range of GPS equipment and software requires the user to be familiar with an ever-expanding body of knowledge.

GPS SEGMENT ORGANIZATION

The Space Segment

Though there has been some evolution in the arrangement, the current GPS constellation under full operational capability consists of 25 usable Block II and IIA satellites and 3 Block IIR satellites. The constellation orbits in six orbital planes, with each plane inclined to the equator by 55° (Figure 3.3). This design covers the globe completely and means that multiple satellite coverage is always available.

Orbital Period

NAVSTAR satellites are more than 20,000 km above the earth in a *posigrade* orbit. A posigrade orbit is one that moves in the same direction as the earth's rotation. Since each satellite is nearly three times the earth's radius above the surface, its orbital period is 12 sidereal hours.

4 Minute Difference

When an observer actually performs a GPS survey project, one of the most noticeable aspects of a satellite's motion is that it returns to the same position in the sky about 4 minutes earlier each day. This apparent regression is attributable to the difference between 24 solar hours and 24 sidereal hours, otherwise known as star-time. GPS satellites retrace the same orbital path twice each sidereal day, but since their observers, on earth, measure time in solar units the orbits do not look quite so regular to them. The satellites actually lose 3 minutes and 56 seconds with each successive solar day.

This rather esoteric fact has practical applications; for example, if the satellites are in a particularly favorable configuration for measurement and the observer wishes to take advantage of the same arrangement the following day, he or she would be well advised to remember the same configuration will occur about 4 minutes earlier on the solar time scale. Both Universal Time (UT) and GPS time are measured in solar, not sidereal units. It is possible that the satellites will be pushed 50 km higher in the future to remove their current 4-minute regression, but for now it remains.

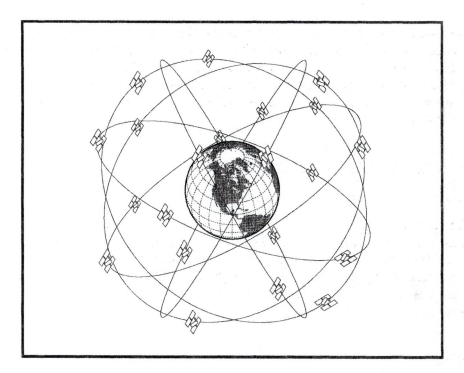

Figure 3.3. Satellite Coverage Constellation.

Design

As mentioned earlier, the GPS constellation was designed to satisfy several critical concerns. Among them were the best possible coverage of the earth with the fewest number of satellites, the reduction of the effects of gravitational and atmospheric drag, sufficient upload and monitoring capability with all control stations located on American soil, and the achievement of maximum accuracy.

Dilution of Precision

The distribution of the satellites above an observer's horizon has a direct bearing on the quality of the position derived from them. Like some of its forerunners, the accuracy of a GPS position is subject to a geometric phenomenon called *dilution of precision (DOP)*. This number is somewhat similar to the strength of figure consideration in the design of a triangulation network. DOP concerns the geometric strength of the figure described by the positions of the satellites with respect to one another and the GPS receivers.

A low DOP factor is good, a high DOP factor is bad. In other words, when the satellites are in the optimal configuration for a reliable GPS position the DOP is low; when they are not, the DOP is high (Figure 3.4).

Four or more satellites must be above the observer's mask angle for the simultaneous solution of the clock offset and three dimensions of the receiver's position.

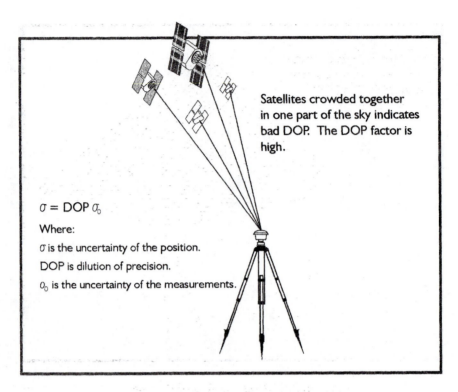

Satellites crowded together in one part of the sky indicates bad DOP. The DOP factor is high.

$\sigma = DOP\ \sigma_0$

Where:

σ is the uncertainty of the position.

DOP is dilution of precision.

σ_0 is the uncertainty of the measurements.

Figure 3.4. Bad Dilution of Precision.

But if all of those satellites are crowded together in one part of the sky, the position would be likely to have an unacceptable uncertainty and the DOP, or dilution of precision, would be high. In other words, a high DOP is like a warning that the actual errors in a GPS position are liable to be larger than you might expect. But remember, it is not the errors themselves that are directly increased by the DOP factor, it is the **uncertainty of the GPS position** that is increased by the DOP factor.

Here is an approximation of the effect:

$$\sigma = DOP\ \sigma_0$$

Where: σ = the uncertainty of the position
          DOP = the dilution of precision factor
          σ_0 = the uncertainty of the measurements

In other words, the standard deviation of the GPS position is the dilution of precision factor multiplied by the standard deviations of the biases in the observables.

Now, since a GPS position is derived from a three-dimensional solution there are several DOP factors used to evaluate the uncertainties in the components of a

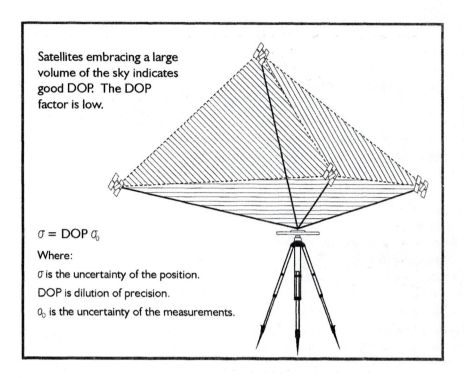

Figure 3.5. Good Dilution of Precision.

GPS position. For example, there is horizontal dilution of precision *(HDOP)* and vertical dilution of precision *(VDOP)* where the uncertainty of a solution for positioning has been isolated into its horizontal and vertical components, respectively. When both horizontal and vertical components are combined, the uncertainty of three-dimensional positions is called *PDOP*, position dilution of precision. There is also *TDOP*, time dilution of precision, that indicates the uncertainty of the clock. There is GDOP, geometric dilution of precision, which is the combination of all of the above. And finally, there is *RDOP*, relative dilution of precision, that includes the number of receivers, the number of satellites they can handle, the length of the observing session as well as the geometry of the satellites' configuration.

The larger the volume of the body defined by the lines from the receiver to the satellites, the better the satellite geometry and the lower the DOP (Figure 3.5). An ideal arrangement of four satellites would be one directly above the receiver, the others 120° from one another in azimuth near the horizon. With that distribution the DOP would be nearly 1, the lowest possible value. In practice, the lowest DOPs are generally around 2. For example, if the standard deviation of a position were ±5 meters and the DOP 2, then the actual **uncertainty** of the position would be 2 times ±5 meters or ±10 meters.

The users of most GPS receivers can set a PDOP mask to guarantee that data will not be logged if the PDOP goes above the set value. A typical PDOP mask is 6.

As the PDOP increases, the accuracy of the positions probably deteriorates, and as it decreases they probably improve, but in neither case is that outcome certain.

Outages

When a DOP factor exceeds a maximum limit in a particular location, indicating an unacceptable level of uncertainty exists over a period of time, that period is known as an *outage*. This expression of uncertainty is useful both in interpreting measured baselines and planning a GPS survey.

Satellite Positions in Mission Planning

The position of the satellites above an observer's horizon is a critical consideration in planning a GPS survey. So, most software packages provide various methods of illustrating the satellite configuration for a particular location over a specified period of time. For example, the configuration of the satellites over the entire span of the observation is important; as the satellites move, the DOP changes. Fortunately the dilution of precision can be worked out in advance. DOP can be predicted. It depends on the orientation of the GPS satellites relative to the GPS receivers. And since most GPS software allows calculation of the satellite constellation from any given position and time, they can also provide the accompanying DOP factors.

Another commonly used plot of the satellite's tracks is constructed on a graphical representation of the half of the celestial sphere. The observer's zenith is shown in the center and the horizon on the perimeter. The program usually draws arcs by connecting the points of the instantaneous azimuths and elevations of the satellites above a specified mask angle. These arcs then represent the paths of the available satellites over the period of time and the place specified by the user.

In Figure 3.6, the plot of the polar coordinates of the available satellites with respect to time and position is just one of several tables and graphs available to help the GPS user visualize the constellation.

Figure 3.7 is another useful graph that is available from many software packages. It shows the correlation between the number of satellites above a specified mask angle and the associated PDOP for a particular location during a particular span of time.

There are four spikes of unacceptable PDOP, labeled here for convenience. It might appear at first glance that these spikes are directly attributable to the drop in the number of available satellites. However, please note that while spike 1 and 4 do indeed occur during periods of 4 satellite data, spikes 2 and 3 are during periods when there are 7 and 5 satellites available, respectively. It is not the number of satellites above the horizon that determine the quality of GPS positions, one must also look at their position relative to the observer, the DOP, among other things. The variety of the tools to help the observer predict satellite visibility underlines the importance of their configuration to successful positioning.

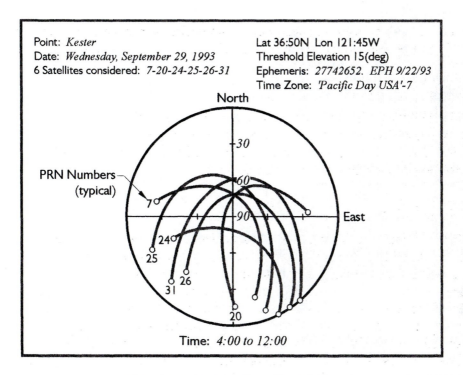

Point: *Kester*
Date: *Wednesday, September 29, 1993*
6 Satellites considered: *7-20-24-25-26-31*

Lat 36:50N Lon 121:45W
Threshold Elevation 15(deg)
Ephemeris: *27742652. EPH 9/22/93*
Time Zone: *'Pacific Day USA'-7*

North

PRN Numbers
(typical)

East

Time: *4:00 to 12:00*

Figure 3.6. Projected Satellite Tracks.

Satellite Names

The 11 GPS satellites launched from Vandenberg Air Force Base between 1978 and 1985 were known as *Block I satellites.* Ten of the satellites built by Rockwell International achieved orbit on Atlas F rockets. There was one launch failure. All were prototype satellites built to validate the concept of GPS positioning. This test constellation of Block I satellites was inclined by 63° to the equator instead of the current specification of 55°. They could be maneuvered by hydrazine thrusters operated by the control stations.

The first GPS satellite was launched February 22, 1978 and was known as *Navstar 1.* An unfortunate complication is that this satellite was also known as PRN 4 just as Navstar 2 was known as PRN 7. The Navstar number, or Mission number includes the Block name and the order of launch; for example I-1, meaning the first satellite of Block I, and the PRN number refers to the weekly segment of the P code that has been assigned to the satellite, and there are still more identifiers. Each GPS satellite has a Space Vehicle number, an Interrange Operation Number, a NASA catalog number, and an orbital position number as well. However, in most literature, and to the GPS receivers themselves, the PRN number is the most important.

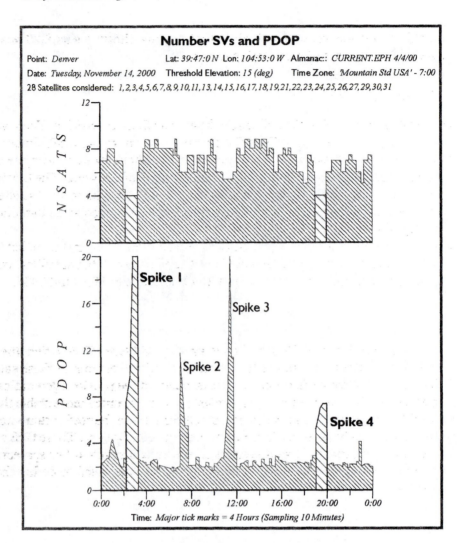

Figure 3.7. Number SVs and PDOP.

Block I

The Block I satellites weighed 845 kg in final orbit. They were powered by three rechargeable nickel-cadmium batteries and 7.25 square meters of single-degree solar panels. These experimental satellites served to point the way for some of the improvements found in subsequent generations. For example, even with the backup systems of two rubidium and two cesium oscillators onboard each satellite, the clocks proved to be the weakest components. The satellites themselves could only store sufficient information for 3.5 days of independent operation. And the uploads from the control segment were not secure; they were not encrypted. Still, all 11 achieved orbit, except NAVSTAR 7. The design life for these satellites was 4.5 years, but their actual average lifetime was 8.76 years. There are no Block I satellites operating today.

Block II and Block IIA

The next generation of GPS satellites are known as *Block II satellites.* There will be 28 of them built. The first left Cape Canaveral on February 14, 1989, almost 14 years after the first GPS satellite was launched. It was about twice as heavy as the first Block I satellite and is expected to have a design life of 7.5 years. The Block II satellites can operate up to 14 days without an upload from the control segment and their uploads are encrypted. The satellites themselves are radiation hardened, and their signals were subject to selective availability.

These satellites were also built by Rockwell International. Block II included the launch of 9 satellites between 1989 and 1990. Nineteen Block IIA satellites, with several navigational improvements, were launched between 1990 and 1997.

Block IIR

The third generation of GPS satellites is known as *Block IIR satellites,* the R stands for *replenishment.* There are two significant advancements in these satellites. First, instead of the cesium and rubidium clocks of the previous generations, these satellites use hydrogen masers. Hydrogen masers are much more stable than earlier oscillators. Second, the Block IIR satellites have enhanced autonomous navigation capability because of their use of intersatellite linkage. These GPS satellites are not only capable of self-navigation, they also provide other spacecraft equipped with an onboard GPS receiver with the data they need to define their own positions.

Block IIF

The fourth generation of GPS satellites is known as *Block IIF satellites;* the F stands for *follow-on.* The program includes the procurement of 33 satellites and the operation and support of a new GPS operations control segment. There is some discussion of including a third civilian frequency on the Block IIF satellites. The frequency may be called L5. The Block IIF satellites are expected to have a design life of 15 years, with the first of them launched around 2005.

Signal Deterioration

While the signals from Block I satellites were not subject to any officially sanctioned deterioration, the same cannot be said of the Block II satellites (Figure 3.8). In the interest of national security the signals from the operational constellation of GPS satellites, including the Block II/IIA and IIR satellites, were intentionally degraded periodically. Selective availability (SA) of the C/A code was implemented

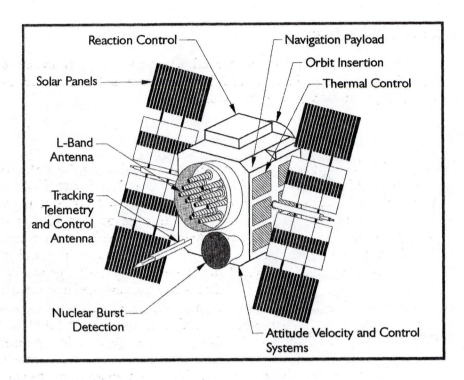

Figure 3.8. GPS Block II Satellite.

by disrupting the satellite clock frequency from time to time from April 1990 until May 2, 2000 at approximately 4:00 UT when it was discontinued.

While SA has been switched off, the P code does continue to be intermittently supplanted by the encrypted Y code in a procedure known as *antispoofing (AS)*. However, this procedure does not significantly affect relative positioning methods that rely on the carrier beat phase observable.

GPS Satellites

All GPS satellites have some common characteristics. They weigh about a ton and with solar panels extended are about 27 feet long. They generate about 700 watts of power. They all have three-dimensional stabilization to ensure that their solar arrays point toward the sun and their 12 helical antennae to the earth. GPS satellites move at a speed of about 8,700 miles per hour. Even so, the satellites must pass through the shadow of the earth from time to time and onboard batteries provide power. All satellites are equipped with thermostatically controlled heaters and reflective insulation to maintain the optimum temperature for the oscillator's operation. Prior to launch, GPS satellites are checked out at a facility at Cape Canaveral, FL.

THE CONTROL SEGMENT

As mentioned in Chapter 1, there are several government tracking and uploading facilities distributed around the world. Taken together these facilities are known as the *Control Segment.* Here is how they work together.

MCS

The Master Control Station (MCS), once located at Vandenberg Air Force Base in California now resides at the Consolidated Space Operations Center (CSOC) at Schriever Air Force Base near Colorado Springs, Colorado and has been manned by the 2nd Space Operations Squadron since 1992. The squadron not only operates and maintains the Master Control Station, but also the network of ground antennas and monitoring stations. The monitoring stations track the navigation signals of all the satellites. That information is then processed at the MCS and is used to update the satellites' navigation messages. The MCS sends updated navigation information to the satellites through ground antennas. The ground antennas are also used to transmit commands to satellites and to receive the satellites' telemetry. This station computes updates for the Navigation message, including the broadcast ephemeris and satellite clock corrections derived from about one week of the tracking information it collects from its five monitoring stations around the world. The MCS also initiates satellite repositioning when it becomes necessary.

Other Stations

The monitoring stations are located at Ascension Island, Colorado Springs, Diego Garcia, Hawaii, and Kwajalein Atoll, with a backup station planned for Onizuka Air Force Base, CA. They track all GPS satellites in view and collect ranging and satellite clock data using the P code pseudoranges and integrated Doppler measurements from all available satellites. Their data are then passed on to MCS over Defense Satellite Communications System satellites. Operators in the MCS calculate each satellite's status, ephemeris, and clock data. This information is sent to antennas located at the monitoring stations (except Hawaii) where the data are uploaded back to each satellite for inclusion in their broadcast messages.

The monitoring stations' measurements of the satellites' actual position are compared with the latest reference ephemeris. The ranges are smoothed by *Kalman filtering* and used in the creation of new estimates of the satellites' position and speed.

Kalman Filtering

Kalman filtering, named for R.E. Kalman's recursive solution for least-squares filtering (Kalman, 1960), has been applied to the results of radionavigation for several decades. It is a statistical method of smoothing and condensing large amounts of data. One of its uses in GPS is reduction of the pseudoranges measured at 1.5-second intervals between a monitoring station and a satellite. Kalman filtering is used to condense a smoothed set of pseudoranges for a 15-minute period that is transmitted to the Master Control Station.

An Analogy

Kalman filtering can be illustrated by the example of an automobile speedometer. Imagine the needle of an automobile's speedometer has a bent cable and is fluctuating between 64 and 72 mph as the car moves down the road. The driver might estimate the actual speed at 68 mph. Although not accepting the speedometer's measurements literally, he has taken them into consideration and constructed an internal model of his velocity. If the driver further depresses the accelerator and the needle responds by moving up, his reliance on the speedometer increases. Despite its vacillation, the needle has reacted as the driver thought it should. It went higher as the car accelerated. This behavior illustrates a predictable correlation between one variable, acceleration, and another, speed. Now he is more confident in his ability to predict the behavior of the speedometer. The driver is illustrating *adaptive gain,* meaning that he is fine-tuning his model as he receives new information about the measurements. As he does, a truer picture of the relationship between the readings from the speedometer and his actual speed emerges, without recording every single number as the needle jumps around. The driver in this analogy is like the Kalman filter.

Without this ability to take the huge amounts of satellite data and condense them into a manageable number of components, GPS processors would be overwhelmed. Kalman filtering is used in the uploading process to reduce the data to the satellite clock offset and drift, 6 orbital parameters, 3 solar radiation pressure parameters, biases of the monitoring station's clock, a model of the tropospheric effect, and earth rotational components.

Constant Tracking

Every GPS satellite is being tracked by at least one of the control segment's monitoring stations at all times. The MCS sends its updates around the world to strategically located uploading stations. They in turn transmit the new Navigation messages to each satellite. The Block II satellites can function without new uploads for 14 days and Block IIA satellites for only 180 days. In any event, the older their Navigation message gets, the more its veracity deteriorates.

Postcomputed Ephemerides

This system is augmented by other tracking networks that produce postcomputed ephemerides. Their impetus have been several: the necessity of timely orbital information with more precision than the broadcast ephemeris and the correlation of the terrestrial coordinate systems with the orbital system through VLBI and SLR sites, among others. The International GPS Geodynamics Service *(IGS)* is an example of a global tracking network. There are also regional organizations such as the network of eight tracking stations which form the Australian Fiducial Network *(AFN)* administered by the Australian Surveying and Land Information Group *(AUSLIG)*.

In the United States, the National Oceanic and Atmospheric Administration *(NOAA)* and, more specifically, the National Geodetic Survey *(NGS)* have been given the job of providing accurate GPS satellite ephemerides, or orbits. Their precise ephemerides are derived using 24-hour data segments from the global GPS network coordinated by the IGS. It is important to note that the former Cooperative International GPS Network or CIGNET is now a part of the IGS.

The IGS consists of 41 continuously operating GPS receivers with more being added to improve the global distribution. Wherever possible, the IGS network receivers are collocated with the very long baseline interferometry (*VLBI*) radio telescopes. The IGS reference frame is the International Earth Rotation Service Terrestrial Reference Frame (*ITRF*).

The precise ephemerides from NGS are available on-line along with a summary file to document the computation and to convey relevant information about the observed satellites, such as maneuvers or maintenance. The precise ephemerides are usually available two to six days after the date of observation. Each data set provides one week's ephemeris information at 15-minute intervals. The on-line address is http://www.ngs.noaa.gov/GPS/GPS.html.

The User Segment

The military plans to build a GPS receiver into virtually all of its ships, aircraft, and terrestrial vehicles. In fact, the Block IIR satellites may be harbingers of the incorporation of more and more receivers into extraterrestrial vehicles as well. But even with such widespread use in the military, civilian GPS will be still more extensive.

Constantly Increasing Application of GPS

The uses the general public finds for GPS will undoubtedly continue to grow as the cost and size of the receivers continues to shrink. The number of users in surveying will be small when compared with the large numbers of trains, cars, boats, and airplanes with GPS receivers. GPS will be used to position all categories of civilian transportation, as well as law enforcement, and emergency vehicles. Nevertheless, surveying and geodesy have the distinction of being the first practical application of GPS and the most sophisticated uses and users are still under its purview. That situation will likely continue for some time.

Next Chapter

The number, range, and complexity of GPS receivers available to surveyors has exploded in recent years. There are widely varying prices and features that sometimes make it difficult to match the equipment with the application. Chapter 4 will be devoted to a detailed discussion of the GPS receivers themselves.

EXERCISES

1. Which GPS satellites carry corner cube reflectors, and what is their purpose?

   (a) SVN 32 and SVN 33 carry onboard corner cubes to allow photo-graphic tracking. The purpose of the reflectors is to allow ground stations to distinguish the satellites, illuminated by earth-based beacons, from the background of fixed stars.
   (b) All GPS satellites carry corner cube reflectors. The purpose of the reflectors is to allow the users to broadcast signals to the satellites that will activate the onboard transponders.
   (c) SVN 36 and SVN 37 carry onboard corner cubes to allow SLR. The purpose of the reflectors is to allow ground stations to distinguish between satellite clock errors and satellite ephemeris errors.
   (d) No GPS satellites carry corner cube reflectors. Such an arrangement would require the user to broadcast a signal from the earth to the satellite. Any requirement that the user reveal his position is not allowed by the military planners responsible for developing GPS. They have always favored a passive system that allowed the user to simply receive the satellite's signal.

2. Which of the following statements concerning the L-band designation is not true?

   (a) The frequency bands used in radar were given letters to preserve military secrecy.
   (b) The GPS carriers, L1 and L2, are named for the L-band radar desig-nation.
   (c) The frequencies broadcast by the TRANSIT satellites were within the L-band.
   (d) The L in L-band stands for long.

3. Which of the following is an aspect of the NAVSTAR GPS system that is an improvement on the retired TRANSIT system?

   (a) rubidium, cesium, and hydrogen maser frequency standards
   (b) satellites that broadcast two frequencies
   (c) a passive system, one that does not require transmissions from the users
   (d) satellites that broadcast their own ephemerides

4. Which of the following requirements for the ideal navigational system, from the military point of view, described in the Army POS/NAV Master Plan in 1990 does GPS not currently satisfy?

(a) The users should be passive.
(b) It should be resistant to countermeasures.
(c) It should be capable of working in real time.
(d) It should not be dependent on externally generated signals.

5. Which of the following statements best explains the fact that for a stationary receiver a GPS satellite appears to return to the same position in the sky about 4 minutes earlier each day?

(a) Over the same period of time VDOP, vertical dilution of precision, is frequently larger than HDOP, horizontal dilution of precision, for a stationary receiver.
(b) The apparent regression is due to the difference between the star-time and solar time over a 24-hour period.
(c) The loss of time is attributable to the satellite's pass through the shadow of the earth.
(d) The apparent regression is due to the cesium and rubidium clocks of earlier GPS satellites. They will be replaced in the Block IIR satellites by hydrogen masers. Hydrogen masers are much more stable than earlier oscillators.

6. All of the following concepts were developed in other contexts and all are now utilized in GPS. Which of them has been around the longest?

(a) orbiting transmitters with accurate frequency standards
(b) Kalman filtering
(c) measuring distances with electromagnetic signals
(d) the Doppler shift

7. Practically speaking, which of the following was the most attractive aspect of first civilian GPS surveying in the early 1980s?

(a) satellite availability
(b) GPS hardware
(c) accuracy
(d) GPS software

8. Which of the following satellite identifiers is most widely used?

(a) Interrange Operation Number
(b) NASA catalog number

    (c)   PRN number

    (d)   NAVSTAR number

9. Satellites in which of the following categories are currently providing signals for positioning and navigation?

    (a)   Block IIR
    (b)   TRANSIT
    (c)   Block IIF
    (d)   Block I

10. Which of the following is not the location of a GPS Control Segment monitoring station?

    (a)   Ascension Island
    (b)   Diego Garcia
    (c)   Cape Canaveral
    (d)   Kwajalein Atoll

ANSWERS AND EXPLANATIONS

1. Answer is (c)

    Explanation: Two current GPS satellites carry onboard corner cube reflectors, SVN 36 (PRN 06) and SVN 37 (PRN 07), launched in 1994 and 1993, respectively. The purpose of the corner cube reflectors is to allow SLR tracking. The ground stations can use this exact range information to separate the effect of errors attributable to satellite clocks from errors in the satellite's ephemerides. Remember the satellite's ephemeris can be likened to a constantly updated coordinate of the satellite's position. The SLR, Satellite Laser Ranging, can nail down the difference between the satellite's broadcast ephemeris and the satellite's actual position. This allows, among other things, more proper attribution of the appropriate portion of the range error to the satellite's clock.

2. Answer is (c)

    Explanation: The original letter designations were assigned to frequency bands in radar to maintain military secrecy. The L-band was given the letter L to indicate that its wavelength was long. The frequencies within the L-band are from 3900 MHz to 1550 MHz, approximately. Stated another way, the wavelengths within the L-band are approximately from 76 cm to 19 cm. The GPS carrier frequencies L1 at 1575.42 MHz and L2 at 1227.60, with wavelengths of 19 cm and 24 cm, respectively, are both close to the L-band range. While one might say that the L1 frequency is not exactly within original L-band designation, the frequencies broadcast by the TRANSIT satellites certainly aren't. The frequencies used in the TRANSIT system are 400 MHz and 150 MHz. These frequencies fall much below the L-band and into the VHF range.

It is interesting to note that the old L-band designation has actually been replaced; it is now known as the D-band. However, there does not appear to be any intention to change the names of the GPS carrier frequencies.

3. Answer is (a)

Explanation: Many of the innovations used in the TRANSIT system informed decisions in creating the NAVSTAR GPS system. Both have satellites that broadcast two frequencies to allow compensation for the ionospheric dispersion. TRANSIT satellites used the frequencies of 400 MHz and 150 MHz, while GPS uses 1575.42 MHz and 1227.60 MHz. And while the low frequencies used in the TRANSIT system were not as effective at eliminating the ionospheric delay, the idea is the same. Both systems are passive, meaning there is no transmission from the user required. Both systems use satellites that broadcast their own ephemerides to the receivers.

And there are more similarities, both systems are divided into three segments: the control segment, including the tracking and upload facilities; the space segment, meaning the satellite themselves; and the user segment, everyone with receivers. And in both the TRANSIT and GPS systems each satellite and receiver contains its own frequency standards. However, the standards used in NAVSTAR GPS satellites are much more sophisticated than those that were used in TRANSIT. The rubidium, cesium, and hydrogen maser frequency standards used in GPS satellites far surpass the quartz oscillators that were used in the TRANSIT satellites. The TRANSIT navigational broadcasts were switched off on December 31, 1996.

4. Answer is (d)

Explanation: Not only does GPS have worldwide 24-hour coverage, it is also capable of providing positions on a huge variety of grids and datums. The system allows the users' receivers to be passive; there is no necessity for them to emit any electronic signal to use the system. GPS codes are complex and there is more than one strategy that the Department of Defense can use to deny GPS to an enemy. At the same time, there can be virtually an unlimited number of users without overtaxing the system. However, GPS is dependent on externally generated signals for the continued health of the system. While satellites can operate for periods without uploads and orbital adjustments from the Control Segment, they certainly cannot do without them entirely.

The Control Segment's network of ground antennas and monitoring stations track the navigation signals of all the satellites. The MCS uses that information to generate updates for the satellites which it uploads through ground antennas.

5. Answer is (b)

   Explanation: The difference between 24 solar hours and 24 sidereal hours, otherwise known as star-time, is 3 minutes and 56 seconds, or about 4 minutes. GPS satellites retrace the same orbital path twice each sidereal day, but since their observers, on earth, measure time in solar units the orbits do not look quite so regular to them, and both Universal Time (UT) and GPS time are measured in solar, not sidereal units.

6. Answer is (d)

   Explanation: The concept of measuring distance with electromagnetic signals had its earliest practical applications in radar in the 1940s and during WWII.

   Development of the technology for launching transmitters with onboard frequency standards into orbit was available soon after the launch of Sputnik in 1957. TRANSIT 1B launched on April 13, 1960 was the first successfully launched navigation satellite.

   Kalman filtering, named for R.E. Kalman's recursive solution for least-squares filtering, was developed in 1960. It is a statistical method of smoothing and condensing large amounts of data and has been used in radionavigation ever since. It is an integral part of GPS.

   However, the Doppler shift was discovered in 1842, and certainly has the longest history of any of the ideas listed. The Doppler shift describes the apparent change in frequency when an observer and a source are in relative motion with respect to one another. If they are moving together the frequency of the signal from the source appears to rise, and if they are moving apart the frequency appears to fall. The Doppler effect came to satellite technology during the tracking of Sputnik in 1957. It occurred to observers at Johns Hopkins University's Applied Physics Laboratory that the Doppler shift of its signal could be used to find the exact moment of its closest approach to the earth. However, the phenomenon had been described 115 years earlier by Christian Doppler using the analogy of a ship on the ocean.

7. Answer is (d)

   Explanation: The interferometric solutions made possible by computerized processing developed with earlier extraterrestrial systems were applied to GPS by the first commercial users, but the software was cumbersome by today's standards. There were few satellites up in the beginning and the necessity of having at least four satellites above the horizon restricted the available observation sessions to difficult periods of time. GPS receivers were large, unwieldy, and very expensive. Nevertheless, in the summer of 1982 a research group at the Massachusetts Institute of Technology (MIT) tested an early GPS receiver and achieved accuracies of 1 and 2 ppm of the station separation. In 1984, a GPS network was produced to control the construction of the Stanford Linear Accelerator. This GPS network provided accuracy at the millimeter level. GPS was inconvenient and expensive, but the accuracy was remarkable from the outset.

8. Answer is (c)

Explanation: The satellite that has currently been given the number Space Vehicle 32 or SV32, is also known as PRN 1. This particular satellite which was launched on November 22, 1992 currently occupies orbital slot F-1. It has a NAVSTAR number or Mission number, of IIA-16. This designation includes the Block name and the order of launch of that mission. This same GPS satellite also has an Interrange Operation Number and a NASA catalog number. However, the most often used identifier for this satellite is PRN 1.

9. Answer is (a)

Explanation: The first of six TRANSIT satellites to reach orbit was launched on June 29, 1961. The constellation of satellites known as the Navy Navigational Satellite System functioned until it was switched off on December 31, 1996, replaced by the GPS system. The first of 11 Block I GPS satellites was launched on February 22, 1978 and the last on October 9, 1985; however, no Block I satellites are operating today. The launch of the first of the GPS satellites known as Block IIF is not expected until 2005.

However, several Block IIR GPS satellites are currently operational. The Block IIR are the third generation of GPS satellites. These satellites use hydrogen masers frequency standards which are much more stable than earlier oscillators. The Block IIR satellites also have enhanced autonomous navigation capability. The first satellite of this category was launched on January 17, 1997.

10. Answer is (c)

Explanation: The monitoring stations are located at Ascension Island, Colorado Springs, Diego Garcia, Hawaii, and Kwajalein Atoll. A backup station is planned for Onizuka Air Force Base, California; however, a Control Segment monitoring station is not located at Cape Canaveral.

The monitoring stations track all GPS satellites in view and collect ranging and satellite clock data using the P code pseudoranges and integrated Doppler measurements. Their data are then passed on to MCS where each satellite's status, ephemeris, and clock error is calculated. This information is sent to antennas located at the monitoring stations (except Hawaii) where the data are uploaded back to each satellite for inclusion in their broadcast messages.

4

Receivers and Methods

COMMON FEATURES OF GPS RECEIVERS

Receivers for GPS Surveying

The most important hardware in a GPS surveying operation are the receivers (Figure 4.1). Their characteristics and capabilities influence the techniques available to the user throughout the work, from the initial planning to processing. There are literally hundreds of different GPS receivers on the market. Only a portion of that number are appropriate for GPS surveying and they share some fundamental elements.

They are generally capable of accuracies from submeter to subcentimeter. They are capable of differential GPS, DGPS, real-time GPS, static GPS, and other hybrid techniques. They usually are accompanied by postprocessing software and network adjustment software. And many are equipped with capacity for extra batteries, external data collectors, external antennas, and tripod mounting hardware. These features, and others, distinguish GPS receivers used in the various aspects of surveying from handheld GPS units designed primarily for recreational use.

A GPS receiver, any GPS receiver, must collect and then convert signals from GPS satellites into measurements. It isn't easy. The GPS signal is only about 40 watts to start with. An orbiting GPS satellite broadcasts this weak signal across a cone of approximately 28° of arc. From the satellite's point of view, about 11,000 miles up, that cone covers the whole planet. It is instructive to contrast this arrangement with a typical communication satellite which not only has much more power, but a very directional signal as well. Its signals are usually collected by a large dish antenna, but the typical GPS receiver has a very small, relatively nondirectional antenna. Fortunately, antennas used for GPS receivers do not even have to be pointed directly at the signal source.

Stated another way, a GPS satellite spreads a low power signal over a large area rather than directing a high power signal at a very specific area. In fact, the GPS signal would be completely obscured by the huge variety of electromagnetic noise that surrounds us if it weren't a spread spectrum coded signal. The GPS signal

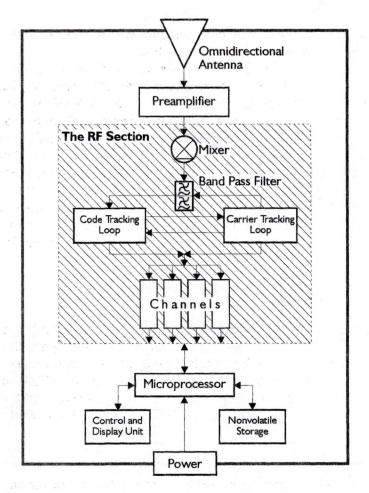

Figure 4.1. A Block Diagram of Code Correlation Receiver.

intentionally occupies a broader frequency bandwidth than it must to carry its information. This characteristic is used to prevent jamming, mitigate multipath, and allow unambiguous satellite tracking.

The Antenna

The antenna collects the satellite's signals. Its main function is the conversion of electromagnetic waves into electric currents sensible to the RF section of the receiver; more about the RF section later. Several antenna designs are possible in GPS, but the satellite's signal has such a low power density, especially after propagating through the atmosphere, that antenna efficiency is critical. Antennas can be designed to collect only the L1 frequency, or both L1 and L2.

Most receivers have an antenna built in, but many can accommodate a separate tripod-mounted or range-pole-mounted antenna as well. These separate antennas with their connecting coaxial cables in standard lengths from 10 to 60 meters are usually available from the receiver manufacturer.

Most of the receiver manufacturers use a *microstrip* antenna. The microstrip can receive one or both GPS frequencies. It is durable, compact, has a simple construction and a low profile. The next most commonly used antenna is known as a *dipole*. A dipole antenna has a stable phase center and simple construction, but needs a good ground plane. A *quadrifilar* antenna is a single frequency antenna. Such antennas have a good gain pattern and do not require a ground plane, but are not azimuthally symmetric. The least common design is the *helix* antenna. A helix is a dual-frequency antenna. It has a good gain pattern, but a high profile. There is also the *choke ring* antenna, built with several concentric rings and designed to reduce the effects of multipath. There is great interest in choke ring antennas both for geodetic surveying and *RTK*, real-time kinematic surveying, even though some tend to be heavy.

Nearly Hemispheric Coverage

The gain pattern of a GPS antenna should provide a nearly full hemisphere of coverage, eliminating any need to aim it at the source of the signal. But the coverage is not usually an absolute hemisphere, partly because most surveying applications filter the signals from very low elevations to reduce the effects of multipath and atmospheric delays. Also, the contours of equal phase around the antenna's electronic center, that is, the *phase center,* are not themselves perfectly spherical.

Antenna Orientation

In an ideal GPS antenna, the phase center of the gain pattern would be exactly coincident with its actual center. If such perfection were possible, the physical centering of the antenna over a point on the earth would ensure its electronic centering as well. But that absolute certainty remains elusive for several reasons.

It is important to remember that the position at each end of a GPS baseline is the position of the phase center of the antenna at each end, not their physical centers. But the phase center is not an immovable point. The location of the phase center actually changes slightly with the satellite's signal. It is different for the L2 signal than it is for the L1 signal. In addition, as the azimuth, intensity, and elevation of the received signal change, so does the difference between the phase center and the physical center. Small azimuthal effects can also be brought on by the local environment around the antenna. But most phase center variation is attributable to changes in satellite elevation. Ignoring these effects can lead to vertical errors to 10 cm.

Fortunately, the errors are systematic when the antennas at each end of a baseline are from the same manufacturer. To compensate for some of this offset error, most receiver manufacturers recommend users take care when making simultaneous observations on a network of points that their antennas are all oriented in the same direction. Several manufacturers even provide reference marks on the antenna so that each one may be rotated to the same azimuth, usually north. That way they are expected to maintain the same relative position between their physical and phase, or electronic, centers when subsequent observations are made.

Height of Instrument

The antenna's configuration also affects another measurement critical to successful GPS surveying—the height of instrument. The measurement of the height of the instrument in a GPS survey is normally made to some reference mark on the antenna. However, it sometimes must include an added correction to bring the total vertical distance to the antenna's phase center.

The Preamplifier

The GPS signal induces a voltage in the antenna that is sent from the receiver to a preamplifier. In fact, a portion of the GPS signal may come from below the mask angle or below the ground plane such that a good deal of it is intentionally not collected by the antenna. A *gain* of about 3 *dB*, decibels, is typical for the usual omnidirectional GPS antenna. The *gain,* or *gain pattern* describes the success of a GPS antenna in collecting more energy from above the mask angle, and less from below the mask angle. The unit of measure used is the decibel, a tenth of a bel. The bel was named for Alexander Graham Bell. Actually, decibel here does not indicate the power of the antenna, it refers to a comparison. In this case, 3 dB indicates that the GPS antenna has about 50% of the capability of an *isotropic* antenna. An isotropic antenna is a hypothetical, lossless antenna that has equal capabilities in all directions.

The antenna is connected to the receiver by a coaxial cable. In the preamplifier the signal's power is increased to endure the attenuation of the cable. The signal is also filtered somewhat to reduce interference before it is sent along into the receiver.

The RF Section

Though different receiver types use different techniques to process the signal, the electronics that do that processing are contained in this section.

Signal processing is easier if the signals arriving from the antenna are lowered to a common frequency band. To accomplish this the incoming frequency is combined with a signal at a harmonic frequency. This latter pure sinusoidal signal is the previously mentioned reference signal generated by the receiver's oscillator. The two frequencies are multiplied together in a device known as a *mixer.* Two frequencies come out, one of them is the sum of the two that went in, and the other is the difference between them.

The two frequencies then go through a *bandpass filter,* this is an electronic filter that removes the unwanted high frequencies and selects the lower of the two. It also eliminates some further noise from the signal. For tracking the P code this filter will have a 20 MHz bandwidth, but it will be around 2 MHz if the C/A code is required. In any case, the signal that results is known as the *IF,* intermediate frequency, or beat frequency signal. This beat frequency is the difference between the Doppler-shifted carrier frequency that came from the satellite and the frequency generated by the receiver's own oscillator.

There are usually several IF stages before copies of it are sent into separate channels, each of which extract the code and carrier information for a particular satellite.

As mentioned before, a replica of the C/A or P code is generated by the receiver's oscillator, and now that is correlated with the noisy IF signal. It is at this point that the pseudorange is measured. Remember, the pseudorange is the time shift required to align the internally generated code with the IF signal, multiplied by the speed of light.

The receiver generates another replica; this time it is a replica of the carrier. That carrier is correlated with the IF signal and the shift in phase can be measured. The continuous phase observable, or observed cycle count, is obtained by counting the elapsed cycles since lock-on and by measuring the fractional part of the phase of the receiver-generated carrier.

Channels

The antenna itself does not sort the information it gathers. The signals from several satellites enter the receiver simultaneously. But in the *channels* of the RF section the undifferentiated signals are identified and segregated from one another.

A channel in a GPS receiver is not unlike a channel in a television set. It is hardware, or a combination of hardware and software, designed to separate one signal from all the others. At any given moment, only one frequency from one satellite can be on one channel at a time. A receiver may have as few as 3 or as many as 40 physical channels. Today, 12 channels is typical. A receiver with 12 channels is also known as a 12 channel *parallel* receiver.

Multiplexing and Sequencing

While a parallel receiver has dedicated separate channels to receive the signals from each satellite that it needs for a solution, a *multiplexing* receiver gathers some data from one satellite for a while and then switches to another satellite and gathers more data and so on. Such a receiver can usually perform this switching quickly enough that it appears to be tracking all of the satellites simultaneously. And when a GPS receiver's channels are not continuously dedicated to just one satellite's signal or one frequency it is either known as a multiplexing or a *sequencing* receiver. These are also known as *fast-switching* or *fast-multiplexing* receivers. A multiplexing receiver must still dedicate one frequency from one satellite to one channel at a time, it just makes that time very short. For example, one channel may be used to track the signal from one satellite for only 20 milliseconds, leave that signal and track another for 20 milliseconds, and then return to the first, or even move on to a third.

In contrast to multiplexing receivers, sequential receivers can seem slow. These receivers also switch from satellite to satellite, but they collect all of the data from one satellite before tracking the next one. So, while a multiplexing receiver moves from satellite to satellite at predetermined intervals, a sequential receiver can get hung up if it is having trouble getting all the necessary information from one satellite in the sequence. Sequential receivers are virtually obsolete.

Even though multiplexing and sequencing receivers are generally less expensive, this strategy of switching channels is used much less today than it was formerly. There are four reasons. While a parallel receiver does not necessarily offer more accurate results, parallel receivers with dedicated channels are faster; a parallel receiver has a more certain phase lock, there is redundancy if a channel fails, and they possess a superior signal-to-noise ratio *(SNR)*.

Tracking Methods

Whether continuous or switching channels are used, a receiver must be able to discriminate between the incoming signals. They may be differentiated by their unique C/A codes on L1, their Doppler shifts, or some other method, but in the end each signal is assigned to its own channel.

Pseudoranging

Today, in most receivers the first procedure in processing an incoming satellite signal is synchronization of the C/A code from the satellite's L1 broadcast, with a replica C/A code generated by the receiver itself. The details of this process, also known as a *code-phase* measurement, are more fully described in Chapter 1. But for the purpose of this discussion, recall that when there is no initial match between the satellite's code and the receiver's replica, the receiver time shifts, or *slews*, the code it is generating until the optimum correlation is found. Then a *code tracking loop,* the delay lock loop, keeps them aligned. The time shift discovered in that process is a measure of the signal's travel time from the satellite to the phase center of the receiver's antenna. Multiplying this time delay by the speed of light gives a range. But it is called a pseudorange in recognition of the fact that it is contaminated by the errors and biases set out in Chapter 2.

An Ambiguity in the C/A Code Pseudorange

There is an ambiguity in a C/A code pseudorange. Unlike the integer ambiguity associated with carrier phase ranging, this ambiguity is rather simply handled by the receiver in the RF section. In any case, please recall from Chapter 1 that the whole C/A code from any particular satellite is repeated every millisecond. Yet it takes from approximately 66 milliseconds to 87 milliseconds for a signal to travel from the orbiting satellite to the receiver. Therefore, there must be 66 to 87 complete and identical C/A code periods in transit between the satellite and the receiver at any given instant.

Stated another way, the C/A code "chipping rate," the rate at which each chip is modulated onto the carrier, is 1.023 Mbps. That means, at light speed, the chip length is approximately 300 m. But the whole C/A code period is 1,023 chips, 1 millisecond long. That is approximately 300 km. And, of course, each satellite repeats its whole 300 km C/A code over and over.

These repeated C/A code periods can be thought of as a number of "rulers" extending from the satellite to the receiver. Each "ruler" is approximately 300 km long with graduations of 300 m. So there would be from about 66 to about 87 of these "rulers" between the satellite and the receiver.

Now, the replica C/A code generated in the receiver can be time-shifted to match one incoming C/A code "ruler" from the satellite, as stated above. But that will only resolve a portion of the time it takes the signal to travel the whole distance from the satellite to the receiver. Unlike the P code, where each satellite transmits a week-long section of the code, and there is no ambiguity between the satellite and receiver, the C/A code is short. In fact, the C/A code is so short that the total transit time of the signal cannot be found directly by code-correlation only, while the raw C/A code observation can initially resolve the range between 0–300 km. This raw observation must still be corrected, by adding the appropriate multiples of 300 km to reach the actual pseudorange. It is a kind of integer ambiguity of "rulers."

A rather glib answer to this problem is: "If the receiver gets the wrong integer, you'll soon know it; 300 km is quite a distance." But a more complete answer goes back to the Navigation message, which you will find discussed briefly in Chapter 1 and in Figure 1.1.

There is a HOW, handover word, near the beginning of every one of the five subframes of the Navigation message. Each of these HOWs contains the Z-count of the first data bit of the subsequent TLM, telemetry word, at the beginning of the very next subframe. Note that this TLM is one of 10 words in a subframe. Like all the other words, it is comprised of 30 data bits, and each data bit is 20 milliseconds long.

The beginning of this 20 millisecond data bit, at the beginning of the TLM word, is perfectly synchronized with the beginning one of the satellite's C/A code periods, one of these "rulers" I've been talking about. This is the start of the resolution of the ambiguity, but since each data bit period is 20 milliseconds long there will be 20 C/A code periods in each one. But it just so happens that at the same instant the X1 count is zero.

Here I must digress with a short, and I'm afraid somewhat incomplete explanation of the X1 subcodes. In any case, the X1 codes are subcodes of the P code. They are generated using four 12-bit shift registers, X1A, X1B, X2A, and X2B. Suffice it to say that an X1 count is exactly as long as 1 P-code chip; that is, 97.75 nanoseconds. Therefore, 10 X1 counts, like 10 P-code chips, are exactly as long as 1 C/A code chip.

It is important to remember that there are 20 C/A code periods in each data bit, as described above. Each of these C/A code periods are made up of 1,023 chips.

And each one of these chips corresponds to 10 sequentially numbered X1 counts. And that sequential count begins with zero. So, at that very instant, the 20 millisecond data bit begins, the TLM word begins, and the C/A code period begins. At the same instant the X1 count is zero.

So, the C/A code ambiguity is resolved by setting the Z count to the HOW value and the X1 count to zero at the beginning of the next subframe of the Navigation message. If the GPS receiver's data bit synchronization is right within 1 millisecond or less, the transit time of the C/A code will be unambiguous and correct. If, on the other hand, the receiver makes an error in its alignment of the 1 millisecond C/A code period with the 20 millisecond data bit, then the X1 count will be out by an integer multiple of 1 millisecond; then a usual technique is to try again with 1 millisecond changes in the X1 count.

Carrier Phase Measurement

Once the receiver has used the Navigation message and measured a pseudorange, it can also read the ephemeris and the almanac information, use GPS time, and, for those receivers that do utilize the P code, use the handover word on every subframe as a stepping stone to tracking the more precise code. But while several manufacturers have now found ways to collect pseudoranges from P-code observables, neither this nor the precision of a C/A code pseudorange alone are adequate for the majority of surveying applications. Therefore, the next step in signal processing for most receivers involves the carrier phase observable.

As stated earlier, just as they produce a replica of the incoming code, most receivers also produce a replica of the incoming carrier wave. And the foundation of carrier phase measurement is the combination of these two frequencies. Remember, the incoming signal from the satellite is subject to an ever-changing Doppler shift, while the replica within the receiver is nominally constant.

Carrier Tracking Loop

The process begins after the PRN code has done its job and the code tracking loop is locked. By mixing the satellite's signal with the replica carrier this process eliminates all the phase modulations, strips the codes from the incoming carrier, and simultaneously creates two intermediate or beat frequencies. As mentioned earlier, one is the sum of the combined frequencies, and the other is the difference. The receiver selects the latter, the difference, with a device known as a *bandpass filter*. Then this signal is sent on to the *carrier tracking loop* where the voltage-controlled oscillator is continuously adjusted to follow the beat frequency exactly.

Doppler Shift

As the satellite passes overhead, the range between the receiver and the satellite changes. That steady change is reflected in a smooth and continuous movement of the phase of the signal coming into the receiver. The rate of that change is reflected in the constant variation of the signal's Doppler shift. But if the receiver's oscillator frequency is matching these variations exactly, as they are happening, it will duplicate the incoming signal's Doppler shift and phase. This strategy of making measurements using the carrier beat phase observable is a matter of counting the elapsed cycles and adding the fractional phase of the receiver's own oscillator. Some of the details of the process are more fully described in Chapter 1.

Range Rate

Doppler information has broad application in signal processing. It can be used to discriminate between the signals from various GPS satellites, to determine integer ambiguities in kinematic surveying (more about that later), as a help in the detection of cycle slips, and as an additional independent observable for autonomous point positioning. But perhaps the most important application of Doppler data is the determination of the *range rate* between a receiver and a satellite (Figure 4.2). Range rate is a term used to mean the rate at which the range between a satellite and a receiver changes over a particular period of time.

With respect to the receiver, the satellite is always in motion, of course, even if the receiver is *static,* meaning stationary. But the receiver may be in motion as well, as it is in kinematic GPS. The ability to determine the instantaneous velocity of a moving vehicle has always been a primary application of GPS and is based on the fact that the Doppler-shift frequency of a satellite's signal is nearly proportional to its range rate.

The Typical Change in the Doppler Shift

To see how it works, let's look at a *static,* that is, stationary, GPS receiver. The signal received would have its maximum Doppler shift, 4.5 to 5 cycles per millisecond, when the satellite is at its maximum range, just as it is rising or setting. But the Doppler shift continuously changes throughout the overhead pass. Immediately after the satellite rises, relative to a particular receiver, its Doppler shift gets smaller and smaller, until the satellite reaches its closest approach. At the instant its radial velocity with respect to the receiver is zero, the Doppler shift of the signal is zero as well. But as the satellite recedes it grows again, negatively, until the Doppler shift once again reaches its maximum extent just as the satellite sets.

Continuously Integrated Doppler

The Doppler-shift and the carrier phase are measured by first combining the received frequencies with the nominally constant reference frequency created by the receiver's oscillator. The difference between the two is the often-mentioned *beat frequency,* an intermediate frequency, and the number of beats over a given time interval is known as the *Doppler count* for that interval. Since the beats can

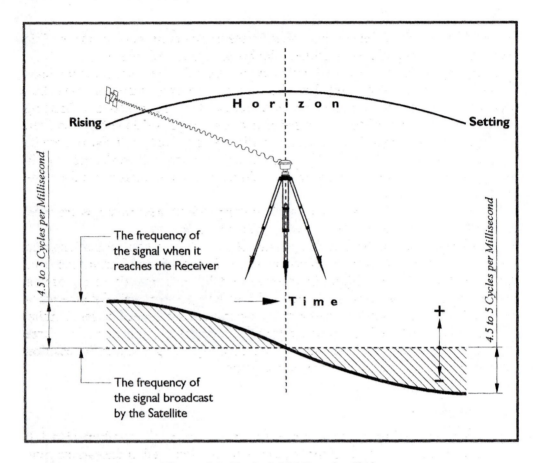

Figure 4.2. Typical GPS Doppler Shift.

be counted much more precisely than their continuously changing frequency can be measured, most GPS receivers just keep track of the accumulated cycles, the Doppler count. The sum of consecutive Doppler counts from an entire satellite pass is often stored, and the data can then be treated like a sequential series of biased range differences. *Continuously integrated Doppler* is such a process. The rate of the change in the continuously integrated Doppler shift of the incoming signal is the same as that of the reconstructed carrier phase.

Integration of the Doppler frequency offset results in an accurate measurement of the advance in carrier phase between epochs. And as stated earlier using double differences in processing the carrier phase observables removes most of the error sources other than multipath and receiver noise. But, as presented in Chapter 2, after the double difference, the integer ambiguity still remains.

Integer Ambiguity

The solution of the integer ambiguity, the number of whole cycles on the path from satellite to receiver, is not easy. But it would be much harder if it were not preceded by pseudoranges, or code phase measurements in most receivers. This allows the centering of the subsequent double-difference solution.

After the code-phase measurements narrow the field there are three methods used to solve the integer ambiguity. The first is a sort of geometric method. The carrier phase data from multiple epochs is processed and the constantly changing satellite geometry is used to find an estimate of the actual position of the receiver. This approach is also used to show the error in the estimate by calculating how its results hold up as the geometry of the constellation changes. It works pretty well, but depends on a significant amount of satellite motion, and therefore, takes time to converge on a solution.

The second approach uses filtering. Here independent measurements are averaged to find the estimated position with the lowest noise level.

The third uses a search through the range of possible integer combinations, and then calculates the one with the lowest residual. The search and filtering methods depend on *heuristic* calculations; in other words, trial and error. These approaches cannot assess the correctness of a particular answer, but can calculate the probability, given certain conditions, that the answer is within a specified set of limits.

In the end, most GPS processing software uses some combination of all three ideas. All of these methods narrow the field by beginning at an initial position estimate provided by code-phase measurements.

Signal Squaring

There is a method that does not use the codes carried by the satellite's signal. It is called codeless tracking, or *signal squaring*. It was first used in the earliest civilian GPS receivers, supplanting proposals for a TRANSIT-like Doppler solution. It makes no use of pseudoranging and relies exclusively on the carrier phase observable. Like other methods it also depends on the creation of an intermediate or beat frequency. But with signal squaring the beat frequency is created by multiplying the incoming carrier by itself. The result has double the frequency and half the wavelength of the original. It is squared.

There are some drawbacks to the method. For example, in the process of squaring the carrier, it is stripped of all its codes. The chips of the P code, the C/A code, and the Navigation message, normally impressed onto the carrier by 180° phase shifts, are eliminated entirely. As discussed in Chapter 1, the signals broadcast by the satellites have phase shifts called *code states* that change from +1 to −1 and vice versa, but squaring the carrier converts them all to exactly 1. The result is that the codes themselves are wiped out. Therefore, this method must acquire information such as almanac data and clock corrections from other sources. Other drawbacks of squaring the carrier include the deterioration of the signal-to-noise ratio because when the carrier is squared the background noise is squared too. And cycle slips occur at twice the original carrier frequency.

But signal squaring has its up side as well. It reduces susceptibility to multipath. It has no dependence on PRN codes and is not hindered by the encryption of the P code. The technique works as well on L2 as it does on L1, and that facilitates dual-frequency ionospheric delay correction. Therefore, signal squaring can provide high accuracy over long baselines.

So there is a cursory look at some of the different techniques used to process the signal in the RF section. Now, on to the microprocessor of the receiver.

The Microprocessor

The microprocessor controls the entire receiver, managing its collection of data. It controls the digital circuits that in turn manage the tracking and measurements, extract the ephemerides, and determine the positions of the satellites, among other things.

The GPS receivers used in surveying often send these data to the storage unit. But more and more they are expected to produce their final positions instantaneously, that is, in *real-time,* as well.

Differential Positioning

There are applications of GPS in which the receiver's microprocessor is expected to provide autonomous single-point positioning using unsmoothed code pseudoranges such as those from inexpensive handheld GPS units. Even though some manufacturers and users make extraordinary claims for their handheld C/A code pseudorange receivers, such autonomous point solutions are not accurate by surveying standards. But even here things have improved. For example, SA's distortion of the satellite clock used to reduce single C/A receiver autonomous positioning to an accuracy of ±100 meters. Since it has been switched off, such positioning can achieve typical accuracies of ±20 to ±40 meters. Still, these are not surveying instruments.

However, code-based pseudoranges using *DGPS*, differential GPS, *can* achieve good real-time, or postprocessed, mapping results. For example, DGPS is often used in collecting data for *Geographical Information Systems, GIS.*

DGPS

The type of differential positioning known as DGPS still depends on code pseudorange observations, but requires at least two receivers. One receiver is placed on a control station and another on an unknown position. They simultaneously track the same codes from the same satellites and because many of the errors in the observations are common to both receivers, the errors are correlated and tend to cancel each other to some degree.

The data from such an arrangement is usually postprocessed, although with a radio link results can be had in real-time. Improvements in this technology have refined the technique's accuracy markedly, and meter- or even submeter results are possible. Still, the positions are not as reliable as those achieved with the carrier phase observable.

Positional Accuracies

With GIS, corner search and mapping work excepted, much GPS surveying requires a higher standard of accuracy. Certainly GPS control surveying often employs several static receivers that simultaneously collect and store data from the same satellites for a period of time known as a *session*. After all the sessions for a day are completed, their data are usually downloaded in a general binary format to the hard disk of a PC for postprocessing.

Kinematic and Real-Time Kinematic

However, not *all* GPS surveying is handled this way. For example, methods such as kinematic surveying, and real-time kinematic surveying also use radio links and the carrier phase observable between GPS receivers. These methods can provide very good results indeed; more about that later.

The CDU

Every GPS receiver has a control and display unit (CDU) and they all have the same fundamental purpose. It allows the operator to query and control the receiver. CDUs come in many shapes and sizes, They range from handheld keyboards to soft keys around a screen. There are digital map displays and interfaces to other instrumentation. But in every case, the CDU is the interface between the operator and the functions of the microprocessor.

Typical Displays

The information available from the CDU varies from receiver to receiver. But when four or more satellites are available they can generally be expected to display the PRN numbers of the satellites being tracked, the receiver's position in three dimensions, and velocity information. Most of them also display the dilution of precision and GPS time.

The Storage

Most GPS receivers today have internal data logging. They use solid state memory or memory cards. Most also allow the user the option of connecting to a computer and having the data downloaded directly to the hard drive. Cassettes, floppy disks, and computer tapes are mostly things of the past in GPS receiver storage.

Downloading

A large number of range measurements and other pertinent data are sent to the receiver's storage during observation sessions. These data are subsequently downloaded through a serial port to a PC or laptop computer to provide the user with the option of postprocessing.

The Power

Battery Power

Receivers operate on battery power generally over a range from 9 to 36 volts DC; most use low voltage. A wide variety of batteries are used, all the way from car batteries to Lithium and Nicad. Some units operate using camcorder batteries. Handheld GPS units sometimes use flashlight batteries.

About half of the available carrier phase receivers have an internal power supply and most will operate 5½ hours or longer on a fully charged 6-amp-hour battery. Most code-tracking receivers, those that do not also use the carrier phase observable, could operate for about 15 hours on the same size battery. There is always a search on for more energy efficient receiver technology.

CHOOSING A GPS RECEIVER

The first question is, what observable is to be tracked? There are receivers that use only the C/A code on the L1 frequency and receivers that use the P code, or encrypted Y code, on both frequencies. There are L1 carrier phase tracking receivers and there are dual-frequency carrier phase tracking receivers.

The C/A code only. Handheld GPS receivers track only the C/A code . Its positions are accurate to approximately ±20 to ±40 meters with SA turned off. Their use by boaters, hikers, even automobile manufacturers is growing. But unlike more sophisticated receivers, they are not capable of measuring the carrier phase observable. These receivers were developed with the needs of navigation in mind. In fact, they are sometimes categorized by the number of waypoints they can store.

Waypoint is a term that grew out of military usage. It means the coordinate of an intermediate position a person, vehicle, or airplane must pass to reach a desired destination. With such a receiver, a navigator may call up a distance and direction from his present location to the next waypoint.

These receivers do not have substantial memory; it isn't needed for their designed applications. However, considerable memory is required for differential GPS, DGPS receivers. It is still a receiver that only tracks the C/A code. But for many applications it must be capable of collecting the same information as is simultaneously collected at a base station on a known point, and storing it for postprocessing. Such an arrangement is capable of meter-level accuracy.

The C/A code and phase on L1. Relative GPS, on lines of less than 10 miles can be accomplished using at least two receivers that track the C/A code and phase of L1. Using double differencing this applications can provide accuracy of approximately ±(1 cm + 2 ppm).

Carrier phase on both frequencies plus the C/A code. A receiver that is capable of observing the carrier phase of both frequencies and the C/A code is appropriate for collecting positions on long baselines, more than 10 miles.

Receiver productivity. Still, choosing the right instrument for a particular application is not easy. Receivers are generally categorized by their physical characteristics, the elements of the GPS signal they can use with advantage, and by the claims about their accuracy. But the effect of these features on a receiver's actual productivity are not always obvious.

For example, it is true that the more aspects of the GPS signal a receiver can employ, the greater its flexibility, but so, too, the greater its cost. And separating the capabilities a user needs to do a particular job from those that are really unnecessary is more complicated than simply listing tracking characteristics, storage capacity, power consumption, and other physical statistics. The user must first have some information about how these features relate to a receiver's performance in particular GPS surveying methods.

Typical concerns. A surveyor generally wants to be able to answer questions like the following. Can this receiver give me the accuracy I need for my work? What is the likely rate of productivity on a project like this? What is the actual cost per point? This section is intended to provide some of the information needed to answer such questions. It offers some general guidelines that are useful in choosing a GPS receiver.

Trends in Receiver Development

In the early years of GPS the military concentrated on testing navigation receivers. But civilians got involved much sooner than expected and took a different direction: receivers with geodetic accuracy.

High Accuracy Early

The first GPS receivers in commercial use were single frequency, six channel, codeless instruments. Their measurements were based on interferometry. As early as the 1980s, those receivers could measure short baselines to millimeter accuracy and long baselines to 1 ppm. It is true the equipment was cumbersome, expensive, and without access to the Navigation message, dependent on external sources for clock and ephemeris information. But they were the first at work in the field and their accuracy was impressive.

Another Direction

During the same era a parallel trend was underway. The idea was to develop a more portable, dual-frequency, four-channel receiver that could use the Navigation message. Such an instrument did not need external sources for clock and ephemeris information, and could be more self-contained.

More Convenience

Unlike the original codeless receivers that required all units on a survey brought together and their clocks synchronized twice a day, these receivers could operate independently. And while the codeless receivers needed to have satellite ephemeris information downloaded before their observations could begin, this receiver could derive its ephemeris directly from the satellite's signal. Despite these advantages, the instruments developed on this model still weighed more than 40 pounds, were very expensive, and were dependent on P-code tracking.

Multichannel and Code-Correlating

A few years later a different kind of multichannel receiver appeared. Instead of the P code, it tracked the C/A code. Instead of using both the L1 and L2 frequencies, it depended on L1 alone. And on that single frequency, it tracked the C/A code and also measured the carrier phase observable. This type of receiver established the basic design for many of the GPS receivers surveyors use today. They are multichannel receivers, and they can recover all of the components of the L1 signal. The C/A code is used to establish the signal lock and initialize the tracking loop. Then the receiver not only reconstructs the carrier wave, it also extracts the satellite clock correction, ephemeris, and other information from the Navigation message. Such receivers are capable of measuring pseudoranges, along with the carrier phase and integrated Doppler observables.

Dual-Frequency

Still, as some of the earlier instruments illustrated, the dual-frequency approach does offer significant advantages. It can be used to limit the effects of ionospheric delay, it can increase the reliability of the results over long baselines, and it certainly increases the scope of GPS procedures available to a surveyor. For these reasons, a substantial number of receivers utilize both frequencies.

Receiver manufacturers are currently using several configurations in building dual-frequency receivers. In order to get both carriers without knowledge of the P code, some use a combination of C/A code-correlation and codeless methods. These receivers use code correlation on L1 and then, borrowing an idea from the first GPS receivers, they add signal squaring on L2.

Adding Codeless Capability

By adding codeless technology on the second frequency, such receivers can avail themselves of the advantages of a dual-frequency capability while avoiding the difficulties of the P code.

Antispoofing, or AS, is the encryption of the P code on both L1 and L2. These encrypted codes are known as the Y codes, Y1 and Y2, respectively. But even though the P code has been encrypted, the carrier phase and pseudorange P code observables have been recovered successfully by many receiver manufacturers.

Adding P-Code Tracking

Dual-frequency receivers are fast becoming the standard for geodetic applications of GPS; some do utilize the P code or the encrypted Y code. There are few receivers with an *entirely* codeless capability. There are no civilian receivers that rely solely on the P code. Nearly all receivers that track the P code use the C/A code on L1. Some use the C/A code on L1, and codeless technology on L2. Some track the P code only when it is not encrypted, and become codeless on L2 when AS is activated. Finally, some track all the available codes on both frequencies. In any of these configurations, the GPS surveying receivers that use the P code in combination with the C/A code and/or codeless technology are among the most expensive.

Typical GPS Surveying Receiver Characteristics

These examples represent the general scope of receivers that provide a level of accuracy acceptable for most surveying applications. Most share some practical characteristics: they have multiple independent channels that track the satellites continuously, they begin acquiring satellites' signals from a few seconds to less than a minute from the moment they are switched on. Most acquire all the satellites above their mask angle in a very few minutes, with the time usually lessened by a warm start, and most provide some sort of audible tone to alert the user that data are being recorded, etc. About three-quarters of them can have their sessions preprogrammed in the office before going to their field sites. And nearly all allow the user to select the *sampling rate* of their phase measurements, from 0 to 3600 seconds.

GPS Receiver Costs

Comparing the cost of GPS receivers over time is complicated by the fact that, today, postprocessing software is often included in their prices. Nevertheless, it is clear that the cost of GPS receivers and GPS technology overall has been through a remarkable decline. The first GPS receivers were five times more expensive than the highest-priced receiver available today. At the same time, the capabilities of even an average receiver have come to outstrip those of the best of the early instruments. These trends will undoubtedly continue, but perhaps the more important aspect of receiver development is the increase in their variety. This growth in diversity has been driven by the rapid expansion in the scope of the uses of GPS.

As previously mentioned, the GPS work done in geodetic and land surveying relies on postprocessed relative positioning. There are several very different techniques available to GPS surveyors. And each method makes unique demands on the receivers used to support it.

In the Future

More and more data processing may find its way into the GPS receivers themselves in the future.

The cost of surveying-grade GPS receivers will probably continue to fall; however, there are some limiting factors to how far that trend will reach. GPS receiver manufacturers have made and continue to make significant investments in GPS software development and the market for surveying receivers is relatively small when compared to the GPS market as a whole.

Multiplexing or fast switching technology is less and less attractive to GPS receiver manufacturers, especially since the current GPS constellation can provide 8 or 10 satellites above the observer's mask angle. Rather, GPS receiver manufacturers are likely to continue to increase the number of channels and opt for continuous tracking of all satellites available.

Surveying-grade GPS receivers are likely to continue to get smaller. This is the logical extension of the ability to pack more electronics into smaller and smaller areas. While power consumption requirements are shrinking too, the big challenge facing GPS receiver manufacturers is the development of long-life miniature batteries.

SOME GPS SURVEYING METHODS

Static

This was the first method of GPS surveying used in the field and it continues to be the primary technique today. Relative static positioning involves several stationary receivers simultaneously collecting data from at least four satellites during observation sessions that usually last from 30 minutes to 2 hours. A typical application of this method would be the determination of vectors, or baselines as they are called, between several static receivers to accuracies from 1 ppm to 0.1 ppm over tens of kilometers (Figure 4.3).

Prerequisites for Static GPS

There are few absolute requirements for relative static positioning. The requisites include: more than one receiver, four or more satellites, and a mostly unobstructed sky above the stations to be occupied. But as in most of surveying, the rest of the elements of the system are dependent on several other considerations.

Productivity

The assessment of the productivity of a GPS survey almost always hinges, in part at least, on the length of the observation sessions required to satisfy the survey specifications. The determination of the session's duration depends on several particulars, such as the length of the baseline and the relative position, that is the geometry, of the satellites among others (Table 4.1).

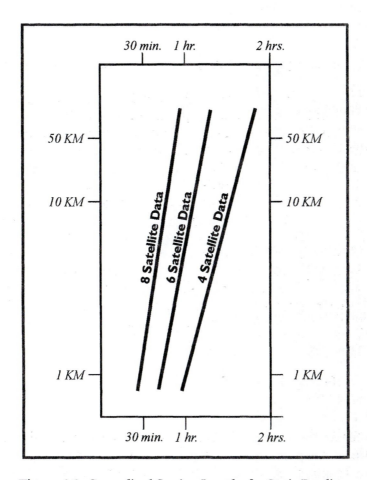

Figure 4.3. Generalized Session Lengths for Static Baselines.

Session Length

Generally speaking, the larger the constellation of satellites, the better the available geometry, the lower the positioning dilution of precision (PDOP), and the shorter the length of the session needed to achieve the required accuracy. For example, given six satellites and good geometry, baselines of 10 km or less might require a session of 45 minutes to 1 hour, whereas, under exactly the same conditions, baselines over 20 km might require a session of 2 hours or more. Alternatively, 45 minutes of six-satellite data may be worth an hour of four-satellite data, depending on the arrangement of the satellites in the sky.

Large Amounts of Data

A static receiver gathers a huge amount of data during an observation session of 1 or 2 hours. Why does it need so much information? The answer may be a bit of a surprise. One might expect there to be some sort of time-consuming process

Table 4.1. Methods

Technique	Accuracy	Observation Time	Drawbacks	Strengths
Static	1/100,000 to 1/5,000,000	1 to 2 hr	Slow.	High accuracy.
Kinematic	1/100,000 to 1/750,000	1 to 2 min	Requires constant lock on at least 4 satellites. Needs initialization.	Fast.
Rapid Static	1/100,000 to 1/1,000,000	5 to 20 min	Requires the most sophisticated equipment.	Very fast and accurate.
On-the-Fly (OTF)	1/100,000 to 1/1,000,000	Virtually instantaneous positioning		Allows highly accurate positions from a receiver in motion.
Pseudokinematic	1/50,000 to 1/500,000	10 to 20 min 2 observations, 1 to 4 hr apart.	Requires 2 separate observations per point.	More productive than static.

that gradually refines the receiver's measurements down to the last millimeter. The situation is quite different. In fact, the receiver actually achieves the millimeter level of accuracy in a matter of seconds, in as little as a single epoch. It is the resolution of the larger divisions of the measurement, the meters, for example, that requires long observation so typical of relative static positioning.

Resolution of the Cycle Ambiguity

The receiver does not need long sessions to make the fine distinctions between millimeters. It needs long sessions to solve the integer number of cycles between itself and the satellites, the so-called *cycle ambiguity* problem. In fact, it is the unique handling of precisely this difficulty that allows the kinematic method to achieve high accuracy with very much shorter occupation times. In the static application, the receivers must resolve the phase ambiguity anew with each occupation, so the sessions are long. But in the kinematic method the receiver resolves the phase ambiguity once, and only once, at the beginning of the project. Then by keeping a continuous lock on the satellite's signals, it maintains that solution throughout the work. This strategy allows for short occupations without sacrificing accuracy. (Please see Chapter 1 for a more thorough presentation of the cause of the cycle ambiguity problem.)

Preprogrammed Observations

Many receivers offer the user facility to preprogram parameters at the occupied station. Some are so automatic they do not require operator interaction at all once they are programmed and on-site, a feature that can be somewhat of a mixed blessing. This will be investigated in more detail in the section on planning a GPS survey.

Observation Settings

The selection of satellites to track, start and stop times, mask elevation angle, assignment of data file names, reference position, bandwidth and sampling rate are some options useful in the static mode, as well as other GPS surveying methods. These features may appear to be prosaic, but their practicality is not always obvious. For example, satellite selection can seem unnecessary when using a receiver with sufficient independent channels to track all satellites above the receiver's horizon without difficulty. However, it is quite useful when the need arises to eliminate data from a satellite that is known to be unhealthy before the observation begins.

Data Interval

Another example is the selectable sampling rate, also known as the *data interval*. This feature allows the user to stipulate the short period of time between each of the microprocessor's downloads to storage. The fastest rate available is usually between 0 and 1 second, and the slowest, 999 and 3600 seconds. The faster the data-sampling rate, the larger the volume of data a receiver collects and the larger the amount of storage it needs. A fast rate is helpful in cycle slip detection, and that improves the receiver's performance on baselines longer than 50 km, where the detection and repair of cycle slips can be particularly difficult.

Compatible Receivers

Relative static positioning, just as all the subsequent surveying methods discussed here, involves several receivers occupying many sites. Problems can be avoided as long as the receivers on a project are compatible. For example, it is helpful if they have the same number of channels and signal processing techniques. And the Receiver Independent Exchange Format, *RINEX,* developed by the Astronomical Institute in 1989, allows different receivers and postprocessing software to work together. Almost all GPS processing software will output RINEX files.

Receiver Capabilities and Baseline Length

The number and type of channels available to a receiver is a consideration because, generally speaking, the more satellites the receiver can track continuously, the better. Another factor that ought to be weighed is whether a receiver has single- or dual-frequency capability. Single-frequency receivers are best applied to relatively short baselines, say, under 25 km. The biases at one end of such a vector are likely to be similar to those at the other. Dual-frequency receivers, on the other hand, have the capability to nearly eliminate the effects of ionospheric refraction, and can handle longer baselines.

Static GPS surveying has been used on control surveys from local to statewide extent, and will probably continue to be the preferred technique in that category.

Differential GPS, DGPS

DGPS is a method that improves GPS pseudorange accuracy. A GPS receiver at a base station, a known position, measures pseudoranges to the same satellites at the same time as other roving GPS receivers. The roving receivers occupy unknown positions in the same geographic area. Occupying a known position, the base station receiver finds corrective factors that can either be communicated in real-time to the roving receivers using a radio link, or may be applied in postprocessing.

Real-time DGPS is somewhat limited to near line-of-sight, whereas postprocessed DGPS can provide meter or submeter positions up to about 300 km from the base station. DGPS is widely used to collect data for GIS applications. There will be more about DGPS later.

Kinematic

Reference Receiver and Rover Receivers

As previously mentioned, kinematic positioning is faster than the static method. The term *kinematic* is sometimes applied to GPS surveying methods where receivers are in continuous motion, but for relative positioning the more typical arrangement is a stop-and-go technique, a method developed by Dr. Benjamin Remondi.

This latter approach involves the use of at least one stationary reference receiver and at least one moving receiver, called a *rover*. The technique is similar to static GPS in that all the receivers observe the same satellites simultaneously, and the reference receivers occupy the same control points throughout the survey. However, the kinematic method differs from static GPS in the movement of the rovers from point to point across the network. They stop momentarily at each new point, usually very briefly, and their data eventually provide vectors between themselves and the reference receivers.

Leapfrog Kinematic

There is a variation of this technique involving two receivers. It is known as the *leapfrog* approach. Instead of either of the receivers remaining a motionless reference throughout the survey, they both move, one at a time. In its simplest form, the receivers take the roving in turns. First receiver A is stationary and receiver B moves; then receiver B waits while receiver A moves. But since they never move at the same time they each take a turn to ensure that at any given moment one of them is providing a static anchor for each vector in the survey, however briefly.

Kinematic Positional Accuracy

When compared with the other relative positioning methods, there is little question that the very short sessions of the kinematic, or real-time kinematic, method can produce the largest number of positions in the least amount of time. The remarkable thing is that this technique can do so with only slight degradation in the accuracy of the work. Recall that in static GPS, the long sessions are not required to resolve the finest aspect of the carrier phase measurement. The millimeters are resolved in a matter of seconds. It is the determination of the meters, the resolution of the integer number of cycles or so-called *phase ambiguity,* that takes the time.

Initialization

The kinematic technique avoids this delay by resolving the phase ambiguity before the survey begins, in a process called *initialization* (Figure 4.4). There are several ways to accomplish initialization. The receivers can occupy each end of a baseline between two control points and since the distance between the points is

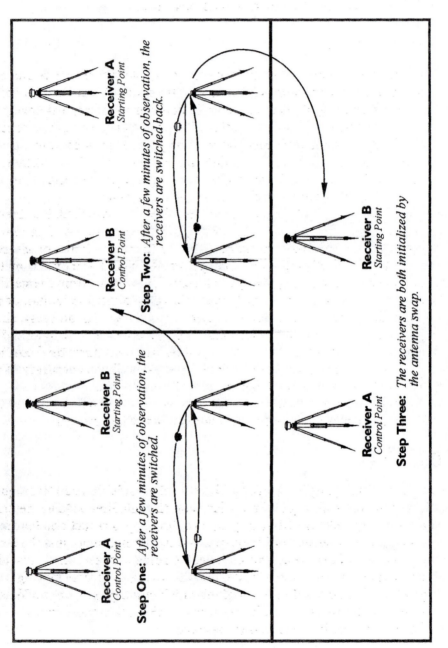

Figure 4.4. The Initialization Process.

known, the phase ambiguity is resolved in a few minutes. Perhaps the best way to establish this baseline is with static GPS techniques. However, the necessary length of the observations means that the speedy kinematic approach starts slow. Kinematic GPS can also be initialized with an *antenna swap.*

Antenna Swap

The antenna swap requires that the two receivers involved are active and collecting data through their antennas from at least four satellites during the entire procedure. While this technique is viable without a control point, it is obviously an advantage if it begins at a known position. One, receiver A, occupies a control point, while the other, receiver B, occupies the first point of the kinematic survey. Obviously, it is also helpful if the two points are conveniently close together. In fact, the distance between receivers can be very small. If it is shorter than the length of their antenna cable the procedure is most convenient.

Following a few minutes of observation the receivers are switched. Receiver A is moved from the control point to the starting point, and receiver B is moved from the starting point to the control point. After another few minutes of observation, they are switched back to their original points. Receiver B is moved from the control point back to the starting point, and receiver A is moved from the starting point back to the control point. With typical tribrach-mounted antennas, or receivers with built-in antennas, the antenna swap procedure is fast and easy, since the tribrachs may stay on their tripods and simply be locked and unlocked to accommodate the instruments. After a few minutes of data in their final position the vector between the two points will be determined to millimeter accuracy. With the receivers both initialized, the kinematic survey may begin.

It is also good practice to return a roving receiver to a known position at the end of a kinematic survey, preferably the station from which it began.

Maintaining Lock

Once the phase ambiguity is resolved with one of these initialization techniques, the receivers are kept running and their locks on the satellites' signals are carefully maintained throughout the survey by avoiding any obstructions that may interrupt them. The drawback of kinematic GPS is that not only must the occupied stations be free of overhead obstructions, the route between them must also be clear, otherwise the lock could be lost while the roving receiver is moving from point to point. In that case, either the original initialization procedure would have to be repeated or the receiver can be returned to the last trouble-free position collected for reinitialization on a known baseline.

Reconnaissance

Therefore, kinematic surveys require considerable reconnaissance. Trees, utility poles, buildings, fences—any sort of structure, natural or man-made—that might prevent the satellite's signals from reaching the receiver for even an instant must be avoided to achieve maximum efficiency with the kinematic method. For this reason it is best to consider five, rather than four, satellites to be the absolute minimum number in planning a kinematic survey. Despite the best reconnaissance, the signal from one satellite may easily be lost during the work. If five are considered the minimum from the outset, such a loss is not so disastrous, since four will remain.

Applications

Kinematic GPS has an ever-widening scope, but it is best suited for work done in wide-open areas. Several kinematic procedures are being developed for airborne and marine work, including photogrammetry without ground control. Land-based kinematic relative positioning applications range from gravity field surveys to construction work. Another example might be a vehicle-mounted receiver moving from station to station along cross-sectional lines. This technique could be used to collect data for a highly accurate topographic map of an area.

Most Carrier Phase Receivers Capable of Kinematic GPS

The majority of the receivers that measure the carrier phase observable are quite adequate to most kinematic applications and generally offer it, along with their static capability. Still, even within that wide selection, some receiver characteristics are more useful than others. For example, a fast data-sampling rate is essential in kinematic surveys. From 1 to 5 seconds is usual, but even faster rates may be needed in work that requires the receivers to be in continuous motion. An option of bandwidth selection in the tracking loops is also helpful in kinematic applications.

Wideband

A wideband signal is transmitted by the GPS satellites because it provides some immunity from jamming and is capable of high-resolution ranging. However, after it has been collected by the receiver, the spread-spectrum signal is usually compressed, or *despread,* into a smaller bandwidth by demodulating it with an identical wideband waveform. Having the option to select the bandwidth is helpful because the narrower it is, the better the signal-to-noise ratio *(SNR).* But there is a tradeoff; although the SNR for the narrower bandwidth is better, a wider bandwidth provides a more secure signal lock. In the end, receivers that can be adapted to find the best balance provide the best results.

Practical Considerations in Kinematic GPS

The physical characteristics of the receiver also need to be considered. When the work is likely to encounter heavy vegetation, a receiver with a separate antenna is a boon, particularly in the kinematic mode. Mounted on a range pole or mast, it can sometimes be elevated above the obstructions.

In kinematic work where the equipment must be packed in, two obvious factors in determining a receiver's suitability are its weight and rate of power consumption. A receiver that provides the user with an audible warning when lock is lost is also quite useful in kinematic applications. This feature may help a user avoid the continuation of a kinematic survey that must be reinitialized. Along the same line, it is a good idea to perform a second antenna swap or another reinitialization procedure at the close of a kinematic survey as a bit of insurance. Then if a loss of lock did occur during the observations, the data may still be rescued by running the postprocessing in reverse, from the closing initialization. There will be more about kinematic GPS later.

Pseudokinematic

This technique has several names. It has come to be known as *pseudokinematic, pseudostatic,* and *intermittent-static*. The variety in its titles is indicative of its standing between static and kinematic in terms of productivity. It is less productive than kinematic, but more productive than the static method. While it is also somewhat less accurate than either relative static or kinematic positioning, its primary advantage is its flexibility. There is no need for the receiver to maintain a lock on the satellite's signals while moving from station to station.

Double Occupation

The pseudokinematic observation procedures can be virtually identical to those previously described for the kinematic relative positioning. However, there is a fundamental difference between the two. In pseudokinematic work, it is necessary to occupy the unknown stations twice. Each one of the two occupations may be as brief, but they should be separated by at least 1 hour and not more than 4 hours. The time between the two occupations allows the satellites of the second session to reach a configuration different enough from that of the first for resolution of the phase ambiguity. This removes the necessity of maintaining a constant lock on the satellite's signals, as required in kinematic work. It also allows the user to occupy many more stations than would be possible with static positioning over the same time.

No Need for Continuous Lock

In pseudokinematic relative positioning, at least two stationary receivers collect up to 10 minutes of data from the same satellites at the same time. A different baseline is occupied by moving one or both of the receivers, without continuous lock. In fact, the receivers may even be switched off during the move. After the move they collect up to 10 minutes of data from the ends of the new baseline. This procedure continues until all of the baselines have been occupied. So far the method is very like other relative positioning techniques, but pseudokinematic differs from the others in that, following the initial occupations, the receivers must return to every one of the baselines in the same manner a second time.

Best Used in Easy Access Situations

While the accuracy of the technique is usually at the subcentimeter level, the production advantage of the procedure over the static mode is somewhat diminished by the reoccupation requirement. Therefore, the best application of pseudokinematic GPS is over relatively short baselines, under 10 km, where the access to the stations is quick and easy.

Mostly Radial Surveys

The majority of receivers that measure the carrier phase observable offer both kinematic and pseudokinematic surveying along with their static capability. Pseudokinematic, as well as kinematic, procedures tend to encourage radial surveys. While the radial approach may be useful for mapping, topography, photocontrol, and some low-order mining surveys, its open-ended quality may not be adequate in other work, like control surveys.

Rapid-Static

Also known as *fast-static,* this field procedure is like pseudokinematic in that it can succeed with very short occupation times. However, it is unlike pseudokinematic in that baselines positioned with rapid-static need be occupied only once. Neither does this procedure require continuous lock on the satellite's signals, as does the kinematic method.

Wide Laning

These advantages are accomplished by relying on the sophisticated hardware of the most expensive receivers. Success in rapid-static depends on the use of receivers that can combine code and carrier phase measurements from both GPS frequencies. The field procedures in rapid-static are very like those of the static method. But using a technique known as *wide laning,* rapid-static can provide the user with nearly the same accuracy available from 1- and 2-hour sessions of relative static positioning with observations of 5 to 20 minutes.

Wide laning is based on the linear combination of the measured phases from both GPS frequencies, L1 and L2. Carrier phase measurements can be made on L1 and L2 separately, of course, but when they are combined two distinct signals result. One is called a *narrow lane* and has a short wavelength of 10.7 cm. The other is known as the *wide lane.* Its frequency, only 347.82 MHz, is more than three times slower than the original carriers. Furthermore, its 86.2 cm wavelength is about four times longer than the 19.0 cm and 24.4 cm of L1 and L2, respectively. These changes greatly increase the spacing of the phase ambiguity, thereby making its resolution much easier.

However, the demands on the receiver are extensive and expensive. It must be capable of dual-frequency tracking and the noise on the signal increases by nearly six times. Nevertheless, wide-laning is very useful for resolving the integer ambiguity very quickly, *on-the-fly, OTF.*

On-the-Fly

Wide laning is also at the heart of a GPS surveying technique called on-the-fly, OTF. This method allows initialization while the receiver is actually in motion.

Accurate Initial Positions

Just as in the rapid-static method, P-code pseudoranges provide for the wide-lane integer bias resolution. That solution is then used, in turn, to solve the base carrier's integer ambiguity. The OTF process differs from rapid static in that the P-code pseudoranges are electronically refined. At the top of the ambiguity resolution process, several consecutive P-code pseudoranges are run through a Kalman filter, a kind of averaging process. This improves the initial position from which the wide-lane and base carrier solutions work, giving the receiver a very accurate, virtually instantaneous idea of where it is. The high-quality starting point makes phase ambiguity resolution very fast.

In fact, these improved P-code pseudoranges offer the user an initialization process that is so fast and accurate it can be accomplished while the receiver is in continuous motion. If the receiver momentarily loses lock on the satellite's signals, with OTF the lock can be reacquired without stopping. Fixing of the integer ambiguities for carrier phase observations on-the-fly is at the heart of precise real-time GPS positioning.

Photogrammetry without Ground Control

The accuracy of OTF is comparable with kinematic GPS, and it lends itself to radial surveys, but it is not crippled by momentary loss of four-satellite data. OTF is a great boon to aerial and hydrographic work and has made the prospect of photogrammetry without ground control a reality, and that is only the beginning of the possible applications of this method.

Real-Time Kinematic, RTK

Like kinematic, here the base station occupies a known position. However, in the real-time kinematic method the base station actually transmits corrections to the roving receiver or receivers using a radio link. The procedure offers high accuracy immediately, in real-time. The results are not postprocessed.

In the earliest use of GPS, kinematic and rapid static positioning were not frequently used because ambiguity resolution methods were still inefficient. But now with integer ambiguity resolution such as on-the-fly (OTF) available, real-time kinematic and similar surveying methods have become very widely used. In fact, today, many surveyors rely heavily on real-time kinematic surveying in their day-to-day work; more about that later.

EXERCISES

1. Which of the following is not a consideration in antenna design for GPS receivers?

   (a)   efficient conversion of electromagnetic waves into electric currents
   (b)   directional capability
   (c)   coincidence of the phase center and the physical center
   (d)   reduction of the effects of multipath

2. Which of the ideas listed below is intended to limit the effect of the difference between the phase center and the physical center of GPS antennas on a baseline measurement?

   (a)   ground plane antennas at each end of the baseline
   (b)   choke ring antennas at each end of the baseline
   (c)   rotation of the antenna's reference marks to north at each end of the baseline
   (d)   the use of the receiver's built-in antennas at each end of the baseline

3. Which of the following does not describe an advantage of a 12-channel parallel continuous-tracking receiver with dedicated channels over a 1-channel multiplexing, or sequencing receiver?

   (a)   The parallel continuous-tracking receiver has a superior signal-to-noise ratio.
   (b)   The parallel continuous-tracking receiver is more accurate.
   (c)   If a channel stops working, there is redundancy with a parallel continuous-tracking receiver.
   (d)   The parallel continuous-tracking receiver has less frequent cycle slips.

4. Which of the following statements concerning the intermediate frequency, IF, in GPS signal processing is correct?

   (a)   In the RF section of a GPS receiver two frequencies go through a *bandpass filter* which selects the higher of the two. This signal is then known as the *IF*, or intermediate frequency.
   (b)   GPS signal processing usually has a single IF stage.

(c) The intermediate frequency is a beat frequency.
(d) The intermediate frequency is the sum of the Doppler-shifted carrier from the satellite and the signal generated by the receiver's own oscillator.

5. Which of the following is not a drawback of signal squaring?

(a) The effect of multipath is increased.
(b) The signal-to-noise ratio deteriorates.
(c) The codes are stripped from the carrier.
(d) The receiver must acquire information such as almanac data and clock corrections somewhere other than the Navigation message.

6. Which of the following is the closest to the length of a C/A code period that is 1,023 chips and 1 millisecond in duration?

(a) 300 meters
(b) 300 kilometers
(c) 66 meters
(d) 20,000 kilometers

7. What is an isotropic antenna?

(a) A hypothetical lossless antenna that has equal capabilities in all directions.
(b) A durable, compact, antenna with a simple construction and a low profile.
(c) An antenna built with several concentric rings and designed to reduce the effects of multipath.
(d) An antenna that has a stable phase center and simple construction, but needs a good ground plane.

8. Which of the following is still used to store data in GPS receivers?

(a) cassette
(b) floppy disk
(c) tapes
(d) internal onboard memory

9. Which of the following two numbers correctly show the proper relationship between the wavelength of the L2 carrier and the wide lane, first, and the increase in the signal noise, second?

(a) about 2 times longer, about 3 times noisier
(b) about 4 times longer, about 6 times noisier

(c) about 10 times slower, about 10 times noisier
(d) 9.23 times faster, 18.16 times noisier

10. Which of the following procedures do kinematic GPS, DGPS, and RTK have in common?

    (a) Each method uses a radio-link between the base station and the rover or rovers.
    (b) Each method uses a base station and rover arrangement.
    (c) Each method must maintain continuous lock on at least four satellites for continuous uninterrupted positioning.
    (d) Each method's results must be postprocessed.

ANSWERS AND EXPLANATIONS

1. Answer is (b)

   Explanation: The GPS signal is actually quite weak and it is broadcast over a very large area. The efficiency of GPS antennas is an important consideration. Limiting the effects of multipath is also a very important consideration. The perfect coincidence of the phase center with the physical center in GPS antennas has yet to be achieved, but it is much sought after and is certainly a design consideration.
   However, the GPS signal is a spread-spectrum coded signal that intentionally occupies a broader frequency bandwidth than it must to carry its information. This fact allows GPS antennas to be omnidirectional. They do not require directional orientation to properly receive the GPS signal.

2. Answer is (c)

   Explanation: Ground-planes, used with many GPS antennas, and the choke ring antenna are designs that attack the same problem, limiting the effects of multipath. The coincidence of the phase center with the physical center in GPS antennas is not yet perfected and since the measurements made by a GPS receiver are made to the phase center of its antenna, its orientation is paramount. The rotation of the antennas at each end of a baseline to north per an imprinted reference is a strategy to reduce the effect of the difference between the two points.

3. Answer is (b)

   Explanation: Multiplexing receivers are at somewhat of a disadvantage when compared with continuous-tracking receivers. Continuous tracking receivers with dedicated channels have a superior signal-to-noise ratio *(SNR)*. They are faster. There is redundancy if a channel fails. They have a more certain phase lock. There are fewer cycle slips.

A multiplexing receiver is not necessarily less accurate than a parallel con-tinuous-tracking receiver.

4. Answer is (c)

Explanation: The intermediate frequency is definitely a beat frequency. It is the combination of the frequency coming from the satellite and a sinusoidal signal generated by the receiver's oscillator. These frequencies are multiplied together in a mixer which produces the sum of the two and the difference. Both go through a bandpass filter which selects the lower of the two and it is known as the intermediate frequency, or beat frequency signal. This beat frequency is the difference between the Doppler-shifted carrier frequency that came from the satellite and the frequency generated by the receiver's own oscillator. There are usually several IF stages in a GPS receiver.

5. Answer is (a)

Explanation: With signal squaring, the beat or intermediate frequency is cre-ated by multiplying the incoming carrier by itself. The result has double the frequency and half the wavelength of the original. It is squared. The process of squaring the carrier strips off all the codes. The P code, the C/A code, and the Navigation message are not available to the receiver. And with the codes wiped out the receiver must acquire information such as almanac data and clock corrections from other sources. Also, the signal-to-noise ratio is degraded, increasing cycle slips. On the other hand, the effect of multipath is actually decreased. And since squaring works as well on L2 as it does on L1, dual-frequency ionospheric delay correction is possible.

6. Answer is (b)

Explanation: The C/A code "chipping rate," the rate at which each chip is modulated onto the carrier, is 1.023 Mbps. That means, at light speed, the chip length is approximately 300 m. But the whole C/A code period is 1,023 chips, 1 millisecond long. That is approximately 300 km. And, of course, each satellite repeats its whole 300 km C/A code over and over.

7. Answer is (a)

Explanation: An isotropic antenna is a hypothetical, lossless antenna that has equal capabilities in all directions. This theoretical antenna is used as a basis of comparison when expressing the gain of GPS, and other antennas. A gain of about 3 dB, decibels, is typical for the usual omnidirectional GPS antenna. The gain, or gain pattern, describes the success of a GPS antenna in collecting more energy from above the mask angle, and less from below the mask angle. The decibel here does not indicate the power of the antenna, it refers to a comparison. In this case, 3 dB indicates that the GPS antenna has about 50 % of the capability of an isotropic antenna.

8. Answer is (d)

Explanation: Most GPS receivers today have internal data logging. They use solid-state memory or memory cards. Most also allow the user the option of connecting to a computer and having the data downloaded directly to the hard drive. Cassette, floppy disks, and computer tapes are mostly things of the past in GPS receiver storage.

9. Answer is (b)

Explanation: Wide laning uses the combination of GPS frequencies, L1 and L2. Carrier phase measurements can be made on L1 and L2 separately, of course, but when they are combined two distinct signals result. One is called a *narrow lane* and has a short wavelength of 10.7 cm. The other is known as the *wide lane.* Its frequency, only 347.82 MHz, is more than three times slower than the original carriers. Furthermore, its 86.2 cm wavelength is about four times longer than the 19.0 cm and 24.4 cm of L1 and L2, respectively, and the noise on the signal increases by nearly six times.

10. Answer is (b)

Explanation: DGPS is a method that improves GPS pseudorange accuracy. A GPS receiver at a base station, usually at a known position, measures pseudoranges to the same satellites at the same time as other roving GPS receivers. The roving receivers occupy unknown positions in the same geographic area. Occupying a known position, the base station receiver finds corrective factors that can either be communicated in real-time to the roving receivers, or may be applied in postprocessing.

   Kinematic GPS is a version of relative positioning in which one receiver is a stationary reference, the base station and at least one other roving receiver coordinates unknown positions with short occupation times while both track the same satellites and maintain constant lock. If lock is lost, reinitialization is necessary to fix the integer ambiguity.

   Real-time kinematic is a method of determining relative positions between known control and unknown positions using carrier phase measurements. A base station at the known position transmits corrections to the roving receiver or receivers. The procedure offers high accuracy immediately, in real-time. The results need not be postprocessed. In the earliest use of GPS, kinematic and rapid static positioning were not frequently used because ambiguity resolution methods were still inefficient. Later, when ambiguity resolution such as on-the-fly (OTF) became available, real-time kinematic and similar surveying methods became more widely used.

5

Coordinates

A FEW PERTINENT IDEAS ABOUT GEODETIC DATUMS FOR GPS

Plane Surveying

Plane surveying has traditionally relied on an imaginary flat reference surface, or *datum,* with Cartesian axes. This rectangular system is used to describe measured positions by ordered pairs, usually expressed in northings and eastings, or *x*- and *y*-coordinates. Even though surveyors have always known that this assumption of a flat earth is fundamentally unrealistic, it provided, and continues to provide, an adequate arrangement for small areas. The attachment of elevations to such horizontal coordinates somewhat acknowledges the topographic irregularity of the earth, but the whole system is always undone by its inherent inaccuracy as surveys grow large.

Development of State Plane Coordinate Systems

In the '30s, an engineer in North Carolina's highway department, George F. Syme, appealed to the then Coast and Geodetic Survey (C&GS, now NGS) for help. He had found that the stretching and compression inevitable in the representation of the curved earth on a plane was so severe over his long-route surveys that he could not check into the C&GS geodetic control stations across his state within reasonable limits. To alleviate the problem, Dr. O.S. Adams of the Division of Geodesy, assisted by Charles N. Claire designed the first state plane coordinate system in 1933. The approach was so successful in North Carolina similar systems were devised for all the states in the Union within a year or so.

The purpose of the state plane coordinate system was to overcome some of the limitations of the horizontal plane datum while avoiding the imposition of geodetic methods and calculations on local surveyors. Using the conic and cylindrical models of the Lambert and Mercator map projections, the flat datum was curved, but only in one direction. By curving the datums and limiting the area of the zones, Dr. Adams managed to limit the distortion to a scale ratio of about 1 part in 10,000 without disturbing the traditional system of ordered pairs of Cartesian coordinates familiar to surveyors.

The state plane coordinate system was a step ahead at that time. To this day, it provides surveyors with a mechanism for coordination of surveying stations that approximates geodetic accuracy more closely than the commonly used methods of small-scale plane surveying. However, the state plane coordinate systems were organized in a time of generally lower accuracy and efficiency in surveying measurement. Its calculations were designed to avoid the lengthy and complicated mathematics of geodesy. It was an understandable compromise in an age when such computation required sharp pencils, logarithmic tables, and lots of midnight oil.

GPS Surveyors and Geodesy

Today, GPS has thrust surveyors into the thick of geodesy which is no longer the exclusive realm of distant experts. Thankfully, in the age of the microcomputer, the computational drudgery can be handled with software packages. Nevertheless, it is unwise to venture into GPS believing that knowledge of the basics of geodesy is, therefore, unnecessary. It is true that GPS would be impossible without computers, but blind reliance on the data they generate eventually leads to disaster.

Some Geodetic Coordinate Systems

3D Cartesian Coordinates

A spatial Cartesian system with three axes lends itself to describing the terrestrial positions derived from space-based geodesy. Using three rectangular coordinates instead of two, one can unambiguously define any position on the earth, or above it for that matter. But such a system is only useful if its origin (0,0,0) and its axes (x,y,z) can be fixed to the planet with certainty, something easier said than done (Figure 5.1).

The usual arrangement is known as the *conventional terrestrial reference system (CTRS),* and the *conventional terrestrial system (CTS).* I will use the latter name. The origin is the center of mass of the earth, the *geocenter.* The x-axis is a line from the geocenter through the intersection of the zero meridian with the equator. The y-axis is extended from the geocenter along a line perpendicular from the x-axis in the same mean equatorial plane. They both rotate with the earth as part of a *right-handed* orthogonal system.

A three-dimensional Cartesian coordinate system is right-handed if it can be described by the following model: the extended forefinger of the right hand symbolizes the positive direction of the x-axis. The middle finger of the same hand extended at right angles to the forefinger symbolizes the positive direction of the y-axis. The extended thumb of the right hand, perpendicular to them both, sym-

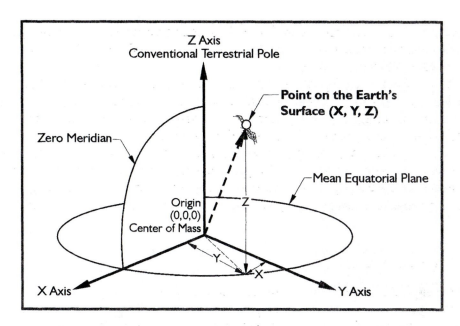

Figure 5.1. Conventional Terrestrial System (CTS);
Earth-Centered-Earth-Fixed Coordinate System (ECEF).

bolizes the positive direction of the z-axis. In applying this model to the earth, the z-axis is imagined to nearly coincide with the earth's axis of rotation, and therein lies a difficulty.

Polar Motion

The earth's rotational axis will not hold still. It actually wanders slightly with respect to the solid earth in a very slow oscillation called *polar motion.* The largest component of the movement relative to the earth's crust has a 430-day cycle known as the *Chandler period.* The actual displacement caused by the wandering generally does not exceed 12 meters. Nevertheless, the conventional terrestrial system of coordinates would be useless if its third axis was constantly wobbling. Therefore, an average stable position was chosen for the position of the pole and the z-axis.

Between 1900 and 1905, the mean position of the earth's rotational pole was designated as the *Conventional International Origin (CIO).* It has since been refined by the *International Earth Rotation Service (IERS)* using very long baseline interferometry (VLBI) and satellite laser ranging (SLR). The name of the z-axis has been changed to the *Conventional Terrestrial Pole (CTP).* But its role has remained the same: also the CTP provides a stable and clear definition on the earth's surface for the z-axis. So, by international agreement, the z-axis of the *Conventional Terrestrial System (CTS)* is a line from the earth's center of mass through the CTP.

The CTS in GPS Postprocessing

The three-dimensional Cartesian coordinates (x,y,z) derived from this system are sometimes known as *earth-centered-earth-fixed (ECEF)* coordinates. They are convenient for many types of calculations in GPS. In fact, most modern GPS software provide data that express vectors as the difference between the *x, y,* and *z* coordinates at each end of the baselines. The display of these differences as *DX, DY,* and *DZ* is a typical product of these postprocessed calculations (Figure 5.2).

Latitude and Longitude

Despite their utility, such 3-D Cartesian coordinates are not the most common method of expressing a geodetic position. Latitude and longitude have been the coordinates of choice for centuries. The application of these angular designations relies on the same two standard lines as 3-D Cartesian coordinates: the mean equator and the zero meridian. Unlike the CTS, they require some clear representation of the terrestrial surface. In modern practice, latitude and longitude cannot be said to uniquely define a position without a clear definition of the earth itself.

Elements of a Geodetic Datum

How can latitude, ϕ, and longitude, λ, be considered inadequate in any way for the definition of a position on the earth? The reference lines—the mean equator and the zero meridian—are clearly defined. The units of degrees, minutes, seconds, and decimals of seconds, allow for the finest distinctions of measurement. Finally, the reference surface is the earth itself.

Despite the certainty of the physical surface of the earth, the *lithosphere,* it remains notoriously difficult to define in mathematical terms. The dilemma is illustrated by the ancient struggle to represent its curved surface on flat maps. There have been a whole variety of map projections developed over the centuries that rely on mathematical relationships between positions on the earth's surface and points on the map. Each projection serves a particular application well, but none of them can represent the earth without distortion. For example, no modern surveyor would presume to promise a client a high-precision control network with data scaled from a map.

As the technology of measurement has improved, the pressure for greater exactness in the definition of the earth's shape has increased. Even with electronic tools that widen the scope and increase the precision of the data, perfection is nowhere in sight.

Development of the Ellipsoidal Model

Despite the fact that local topography is the most obvious feature of the lithosphere to an observer standing on the earth, efforts to grasp the more general nature of the planet's shape and size have been occupying scientists for at least

Start Time:	12.14/93 23:24:45 GPS	(727 257085)
Stop Time:	12/14/93 23:55:15 GPS	(727 258915)
Occupation Time:	00:30:30:00	
Measurement Epoch Interval (seconds):	15:00	

Solution Time: Receiver/Satellite double difference
Iono free fixed

Solution Acceptability: Passed ratio test
Baseline Slope Distance   Std. Dev. (meters): 17044.376   0.000574

	Forward	Backward
Normal Section Azimuth:	179^0 04′ 38.886169″	359^0 04′ 47.857511″
	0^0 07′ 00.456589″	-0^0 16′ 12.548396″

Baseline Components (meters):

dn	-17042.131	de	274.423	du	34.744

Standard Deviations:

dx	-6552.297	dy	-10264.496	dz	-11925.530
	8.437168E-004		1.072974E-003		9.513724E-004

Aposteriori Covariance Matrix:

$$
\begin{matrix}
7.118580\text{E-}007 & & \\
7.389287\text{E-}007 & 1.151273\text{E-}006 & \\
-6.141036\text{E-}007 & -7.690377\text{E-}007 & 9.051094\text{E-}004
\end{matrix}
$$

Variance Ratio   Cutoff: 62.1   1.5
Reference Variance: 0.556

Observable   Count/Rejected   RMS: Iono free phase   451/0   0.006

Figure 5.2. Differences as DX, DY, and DZ.

2,300 years. There have, of course, been long intervening periods of unmitigated nonsense on the subject. Ever since 200 B.C. when Eratosthenes almost calculated the planet's circumference correctly, geodesy has been getting ever closer to expressing the actual shape of the earth in numerical terms. A leap forward occurred with Newton's thesis that the earth was an ellipsoid rather than a sphere in the first edition of his *Principia* in 1687.

Newton's idea that the actual shape of the earth was slightly ellipsoidal was not entirely independent. There had already been some other suggestive observations. For example, 15 years earlier astronomer J. Richter had found that to maintain the accuracy of the one-second clock he used in his observations in Cayenne, French Guiana, he had to shorten its pendulum significantly. The clock's pendulum, regulated in Paris, tended to swing more slowly as it approached the equator. Newton reasoned that the phenomenon was attributable to a lessening of the force of gravity. Based on his own theoretical work, he explained the weaker gravity by the proposition, "the earth is higher under the equator than at the poles, and that by an excess of about 17 miles" (*Philosophiae naturalis principia mathematica,* Book III, Proposition XX).

Although Newton's model of the planet bulging along the equator and flattened at the poles was supported by some of his contemporaries, notably Huygens, the inventor of Richter's clock, it was attacked by others. The director of the Paris Observatory, Jean Dominique Cassini, for example, took exception to Newton's concept. Even though the elder Cassini had himself observed the flattening of the poles of Jupiter in 1666, neither he nor his equally learned son Jacques were prepared to accept the same idea when it came to the shape of the earth. It appeared they had some empirical evidence on their side.

For geometric verification of the earth model, scientists had employed arc measurements at various latitudes since the early 1500s. Establishing the latitude of their beginning and ending points astronomically, they measured a cardinal line to discover the length of one degree of longitude along a meridian arc. Early attempts assumed a spherical earth and the results were used to estimate its radius by simple multiplication. In fact, one of the most accurate of the measurements of this type, begun in 1669 by the French abbé J. Picard, was actually used by Newton in formulating his own law of gravitation. However, Cassini noted that close analysis of Picard's arc measurement, and others, seemed to show the length of one degree of longitude actually *decreased* as it proceeded northward. He concluded that the earth was not flattened as proposed by Newton, but was rather elongated at the poles.

The argument was not resolved until two expeditions between about 1733 and 1744 were completed. They were sponsored by the Paris Académie Royale des Sciences and produced irrefutable proof. One group which included Clairaut and Maupertuis was sent to measure a meridian arc near the Arctic Circle, 66°20′ Nϕ, in Lapland. Another expedition with Bouguer and Godin, to what is now Ecuador, measured an arc near the equator, 01°31′ Sϕ. Newton's conjecture was proved correct, and the contradictory evidence of Picard's arc was charged to errors in the latter's measurement of the astronomic latitudes.

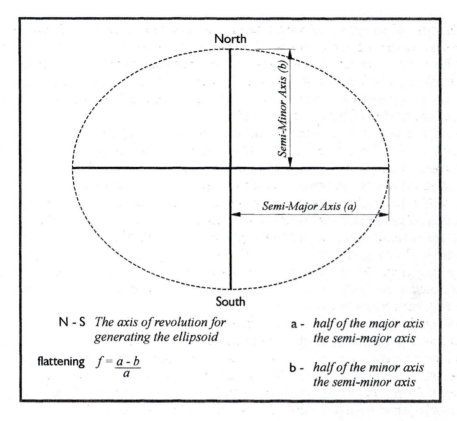

Figure 5.3. Ellipsoidal Model of Earth.

The ellipsoidal model (Figure 5.3), bulging at the equator and flattened at the poles, has been used ever since as a representation of the general shape of the earth's surface. It is called an oblate spheroid. In fact, several reference ellipsoids have been established for various regions of the planet. They are precisely defined by their semimajor axis and flattening. The relationship between these parameters are expressed in the formula:

$$f = \frac{a-b}{a}$$

Where f = flattening
      a = semimajor axis
      b = semiminor axis.

The Role of an Ellipsoid in a Datum

The semimajor axis and flattening can be used to completely define an ellipsoid of revolution. The ellipsoid is revolved around the minor access. However, six additional elements are required if that ellipsoid is to be used as a *geodetic datum:* three to specify its center and three more to clearly indicate its orientation around that center. The *Clarke 1866 spheroid* is one of many reference ellipsoids. Its shape is completely defined by a semimajor axis, *a*, of 6378.2064 km and a flattening, *f*, of 1/294.9786982. It is the reference ellipsoid of the datum known to surveyors as the *North American Datum of 1927 (NAD27)*, but it is not the datum itself.

For the Clarke 1866 spheroid to become NAD27, it had to be attached at a point and specifically oriented to the actual surface of the earth. However, even this ellipsoid, which fits North America best of all, could not conform to that surface perfectly. Therefore, the initial point was chosen near the center of the antici- pated geodetic network to best distribute the inevitable distortion. The attach- ment was established at Meades Ranch, Kansas, 39°13′26″.686 Nϕ, 98°32′30″.506 Wλ and *geoidal height* zero (we will discuss geoidal height later). Those coordi- nates were not sufficient, however. The establishment of directions from this ini- tial point was required to complete the orientation. The azimuth from Meades Ranch to station Waldo was fixed at 75°28′09″.64 and the deflection of the verti- cal set at zero (more later about the deflection of the vertical).

Once the initial point and directions were fixed, the whole orientation of NAD27 was established, including the center of the reference ellipsoid. Its center was imagined to reside somewhere around the center of mass of the earth. However, the two points were certainly not coincident, nor were they intended to be. In short, NAD27 does not use a geocentric ellipsoid.

Measurement Technology and Datum Selection

In the period before space-based geodesy was tenable, a regional datum was not unusual. The *Australian Geodetic Datum 1966*, the *European Datum 1950*, and the *South American Datum 1969*, among others, were also designed as nongeocentric systems. Achievement of the minimum distortion over a particular region was the primary consideration in choosing their ellipsoids, not the relationship of their centers to the center of mass of the earth (Figure 5.4). For example, in the Con- ventional Terrestrial System (CTS) the 3-D Cartesian coordinates of the center of the Clarke 1866 spheroid as it was used for NAD27 are about X = –4 m, Y = +166 m and Z = +183 m.

This approach to the design of datums was bolstered by the fact that the vast majority of geodetic measurements they would be expected to support were of the classical variety. That is, the work was done with theodolites, towers, and tapes. They were earthbound. Even after the advent of electronic distance measurement, the general approach involved the determination of horizontal coordinates by measuring from point to point on the earth's surface and adding heights, other- wise known as *elevations*, through a separate leveling operation. As long as this methodological separation existed between the horizontal and vertical coordinates of a station, the difference between the ellipsoid and the true earth's surface was not an overriding concern. Such circumstances did not require a geocentric da- tum.

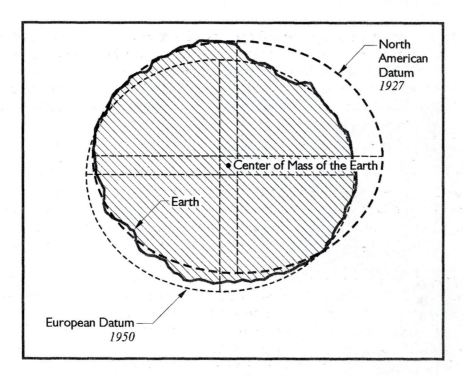

Figure 5.4. Regional Datums.

However, as the sophistication of satellite geodesy increased, the need for a truly global, geocentric datum became obvious. The horizontal and vertical information were no longer separate. Since satellites orbit around the center of mass of the earth, a position derived from space-based geodesy can be visualized as a vector originating from that point.

So, today, not only are the horizontal and vertical components of a position derived from precisely the same vector, the choice of the coordinate system used to express them is actually a matter of convenience. The position vector can be transformed into the 3D Cartesian system of CTS, the traditional latitude, longitude, and height, or virtually any other well-defined coordinate system. However, since the orbital motion and the subsequent position vector derived from satellite geodesy are themselves earth-centered, it follows that the most straightforward representations of that data are earth-centered as well (Figure 5.5).

The Development of a Geocentric Model

Satellites have not only provided the impetus for a geocentric datum, they have also supplied the means to achieve it. In fact, the orbital perturbations of man-made near-earth satellites have probably brought more refinements to the understanding of the shape of the earth in a shorter span of time than was ever before possible. For example, the analysis of the precession of Sputnik 2 in the late '50s showed researchers that the earth's semiminor axis was actually 85 meters shorter than had

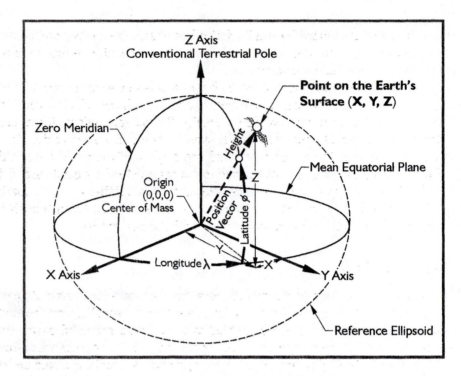

Figure 5.5. A Position Derived from Space-Based Geodesy Visualized as a Vector Originating from the Earth's Center.

been previously thought. In 1958, while studying the tracking data from the orbit of Vanguard I, Ann Bailey of the Goddard Spaceflight Center discovered that the planet is shaped a bit like a pear. There is a slight protuberance at the North Pole, a little depression at the South Pole, and a small bulge just south of the equator.

These formations and others have been discovered through the observation of small distortions in satellites' otherwise elliptical orbits, little bumps in their road, so to speak. The deviations are caused by the action of earth's gravity on the satellites as they travel through space. Just as Richter's clock reacted to the lessening of gravity at the equator and thereby revealed one of the largest features of the earth's shape to Newton, small perturbations in the orbits of satellites, also responding to gravity, reveal details of the earth's shape to today's scientists. The common aspect of these examples is the direct relationship between direction and magnitude of gravity and the planet's form. In fact, the surface that best fits the earth's gravity field has been given a name. It is called the *geoid*.

The Geoid

An often-used description of the geoidal surface involves idealized oceans. Imagine the oceans of the world utterly still, completely free of currents, tides, friction, variations in temperature, and all other physical forces, except gravity. Reacting to gravity alone, these unattainable calm waters would coincide with the figure known as the geoid. Admitted by small frictionless channels or tubes and allowed to migrate across the land, the water would then, theoretically, define the same geoidal surface across the continents, too.

Of course, the 70% of the earth covered by oceans is not so cooperative, nor is there any such system of channels and tubes. In addition, the physical forces eliminated from the model cannot be avoided in reality. These unavoidable forces actually cause mean sea level to deviate up to 1, even 2, meters from the geoid. This is one of the reasons that Mean Sea Level and the surface of the geoid are not the same. And it is a fact frequently mentioned to emphasize the inconsistency of the original definition of the geoid as it was offered by J.B. Listing in 1872. Listing thought of the geoidal surface as equivalent to mean sea level. Even though his idea does not stand up to scrutiny today, it can still be instructive.

An Equipotential Surface

Gravity is not consistent across the topographic surface of the earth. At every point it has a magnitude and a direction. In other words, anywhere on the earth, gravity can be described by a mathematical vector. Along the solid earth, such vectors do not have all the *same* direction or magnitude, but one can imagine a surface of constant gravity potential. Such an *equipotential* surface would be *level* in the true sense. It would coincide with the top of the hypothetical water in the previous example. Despite the fact that real mean sea level does not define such a figure, the geoidal surface is not just a product of imagination. For example, the vertical axis of any properly leveled surveying instrument and the string of any stable plumb bob are perpendicular to the geoid. Just as pendulum clocks and earth-orbiting satellites, they clearly show that the geoid is a reality.

Geoidal Undulation

Just as the geoid does not precisely follow mean sea level, neither does it exactly correspond with the topography of the dry land. It is irregular like the terrestrial surface, and it has similar peaks and valleys. It is bumpy. Uneven distribution of the mass of the planet makes it maddeningly so. Maddening because if the solid earth had no internal anomalies of density, the geoid would be smooth and almost exactly ellipsoidal. In that case, the reference ellipsoid could fit the geoid to near perfection and the lives of geodesists would be much simpler. But like the earth itself, the geoid defies such mathematical consistency and departs from true ellipsoidal form by as much as 100 meters in places (Figure 5.6).

The Modern Geocentric Datum

Three distinct figures are involved in a geodetic datum for latitude, longitude, and height: the geoid, the reference ellipsoid, and the earth itself. Due in large measure to the ascendancy of satellite geodesy, it has become highly desirable that they share a common center.

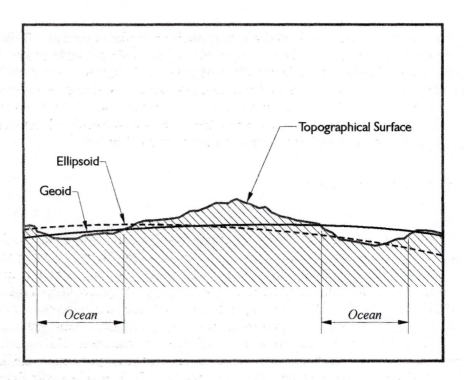

Figure 5.6. The Geoid Departs from True Ellipsoidal Form in Places.

While the level surface of the geoid provides a solid foundation for the definitions of heights (more about that later) and the topographic surface of the Earth is necessarily where measurements are made, neither can serve as the reference surface for geodetic positions. From the continents to the floors of the oceans, the solid earth's actual surface is too irregular to be represented by a simple mathematical statement. The geoid, which is sometimes under, and sometimes above, the surface of the earth, has an overall shape that also defies any concise geometrical definition. But the ellipsoid not only has the same general shape as the earth, but, unlike the other two figures can be described simply and completely in mathematical terms.

Therefore, a global geocentric system has been developed based on the ellipsoid adopted by the *International Union of Geodesy and Geophysics (IUGG)* in 1979. It is called the *Geodetic Reference System 1980 (GRS80)*. Its semimajor axis, *a*, is 6378.137 km and is probably within a few meters of the earth's actual equatorial radius. Its flattening, *f*, is 1/298.25722 and likely deviates only slightly from the true value, a considerable improvement over Newton's calculation of a flattening ratio of 1/230. But then he did not have orbital data from near-earth satellites to check his work.

With very slight changes, GRS80 is the reference ellipsoid for the coordinate system, known as the *World Geodetic System 1984 (WGS84)*. This datum has been used by the U.S. military since January 21, 1987, as the basis for the GPS Navigation message computations. Therefore, coordinates provided directly by GPS receivers are based in WGS84. However, most available GPS software can transform those coordinates to a number of other datums as well. The one that is probably of greatest interest to surveyors in the United States today is the *North American Datum 1983 (NAD83)*. But the difference between WGS84 and NAD83 coordinates is so small, usually about the 0.1-mm level, that transformation is unnecessary and they can be considered equivalent for most applications.

North American Datum 1983

NAD27

The Clarke 1866 ellipsoid was the foundation of NAD27, and the blocks that built that foundation were made by geodetic triangulation. After all, an ellipsoid, even one with a clearly stated orientation to the earth, is only an abstraction until physical, identifiable control stations are available for its practical application. During the tenure of NAD27, control positions were tied together by tens of thousands of miles of triangulation and some traverses. Its measurements grew into chains of figures from Canada to Mexico and coast to coast, with their vertices perpetuated by bronze disks set in stone, concrete, and other permanent media.

These tri-stations, also known as brass caps, and their attached coordinates have provided a framework for all types of surveying and mapping projects for many years. They have served to locate international, state, and county boundaries. They have provided geodetic control for the planning of national and local projects, development of natural resources, national defense, and land management. They have enabled surveys to be fitted together, provided checks, and assisted in the perpetuation of their marks. They have supported scientific inquiry, including crustal monitoring studies and other geophysical research. But even as application of the nationwide control network grew, the revelations of local distortions in NAD27 were reaching unacceptable levels.

Judged by the standards of newer measurement technologies, the quality of some of the observations used in the datum were too low. That, and its lack of an internationally viable geocentric ellipsoid, finally drove its positions to obsolescence. The monuments remain, but it was clear early on that NAD27 had some difficulties. There were problems from too few baselines, Laplace azimuths, and other deficiencies. By the early 1970s the NAD27 coordinates of the national geodetic control network were no longer adequate.

The Development of NAD83

While a committee of the National Academy of Sciences advocated the need for a new adjustment in its 1971 report, work on the new datum, NAD83, did not really begin until after July 1, 1974. Leading the charge was an old agency with a new name. Called the *U.S. Coast & Geodetic Survey* in 1878, and then the *Coast and Geodetic Survey (C&GS)* from 1899, the agency is now known as the *National*

Geodetic Survey (NGS). It is within the *National Oceanic and Atmospheric Administration (NOAA)*. The first ancestor of today's NGS was established back in 1807 and was known as the *Survey of the Coast*. Its current authority is contained in United States Code, Title 33, USC 883a.

NAD83 includes not only the United States, but also Central America, Canada, Greenland, and Mexico. The NGS and the Geodetic Survey of Canada set about the task of attaching and orienting the GRS80 ellipsoid to the actual surface of the earth, as it was defined by the best positions available at the time. It took more than 10 years to readjust and redefine the horizontal coordinate system of North America into what is now NAD83. More than 1.7 million weighted observations derived from classical surveying techniques throughout the Western Hemisphere were involved in the least-squares adjustment. They were supplemented by approximately 30,000 EDM-measured baselines, 5,000 astronomic azimuths, and more than 650 Doppler stations positioned by the TRANSIT satellite system. Over 100 Very Long Baseline Interferometry (VLBI) vectors were also included. But GPS, in its infancy, contributed only five points.

GPS was growing up in the early '80s and some of the agencies involved in its development decided to join forces. NOAA, the *National Aeronautics and Space Administration (NASA)*, the *United States Geological Survey (USGS)*, and the Department of Defense coordinated their efforts. As a result each agency was assigned specific responsibilities. NGS was charged with the development of specifications for GPS operations, investigation of related technologies, and the use of GPS for modeling crustal motion. It was also authorized to conduct its subsequent geodetic control surveys with GPS. So, despite an initial sparseness of GPS data in the creation of NAD83, the stage was set for a systematic infusion of its positions as the datum matured. The work was officially completed on July 31, 1986.

The International Terrestrial Reference System (ITRS)

As NAD83 has aged there has been constant improvement in geodesy. When NAD83 was created it was intended to be geocentric. It is now known that NAD83 is about 2 meters from the true geocenter. Other difficulties have arisen, such as the limitations of the Doppler observations used in the definition of NAD83. Please recall that GPS only contributed five positions to the original work.

The best geocentric reference frame currently available is the International Terrestrial Reference Frame, *ITRF*. Its origin is at the center of mass of the whole earth including the oceans and atmosphere. The unit of length is the meter. The orientation of its axes was established as consistent with that of the IERS's predecessor, Bureau International de l'Heure, BIH, at the beginning of 1984.

Today, the ITRF is maintained by the International Earth Rotation Service, IERS, which monitors Earth Orientation Parameters, EOP, for the scientific community through a global network of observing stations. This is done with GPS, Very Long Baseline Interferometry, *VLBI*, Lunar Laser Ranging, *LLR*, Satellite Laser Ranging, *SLR*, and Doppler Orbitography and Radiopositioning Integrated by Satellite (DORIS) and the positions of the observing stations are now considered to be accurate to the centimeter level.

The International Terrestrial Reference Frame (ITRF) is actually revised and published on a regular basis. And today NAD83 can be defined in terms of a best-fit transformation from ITRF96.

The Management of NAD83

With the surveying capability of GPS and the new NAD83 reference system in place, NGS began the long process of a nationwide upgrade of their control networks. Now known as the *National Geodetic Reference System (NGRS)*, it actually includes three networks. A horizontal network provides geodetic latitudes and longitudes in the North American Datums. A vertical network furnishes heights, also known as elevations, in the *National American Vertical Datum 1988 (NAVD88)*. A gravity network supplies gravity values in the U.S. absolute gravity reference system. Any particular station may have its position defined in one, two, or all three networks.

NGS is computing and publishing NAD83 values for monumented stations, old and new, throughout the United States. Gradually, the new information will provide the common-coordinate basis that is so important to all surveying and mapping activities. But the pace of such a major overhaul must be deliberate, and a significant number of stations will still have only NAD27 positions for some time to come. This unevenness in the upgrade from NAD27 to NAD83 causes a recurrent problem to GPS surveyors across the country.

Since geodetic accuracy with GPS depends on relative positioning, surveyors continue to rely on NGS stations to control their work just as they have for generations. Today, it is not unusual for surveyors to find that some NGS stations have published coordinates in NAD83 and others, perhaps needed to control the same project, only have positions in NAD27. In such a situation it is often desirable to transform the NAD27 positions into coordinates of the newer datum. But, unfortunately, there is no single-step mathematical approach that can do it accurately.

The distortions between the original NAD27 positions are part of the difficulty. The older coordinates were sometimes in error as much as 1 part in 15,000. Problems stemming from the deflection of the vertical, lack of correction for geoidal undulations, low-quality measurements, and other sources contributed to inaccuracies in some NAD27 coordinates that cannot be corrected by simply transforming them into another datum.

Transformations from NAD27 to NAD83

Nevertheless, various approximate methods are used to transform NAD27 coordinates into supposed NAD83 values. For example, the computation of a constant local translation is sometimes attempted using stations with coordinates in both systems as a guide. Another technique is the calculation of two translations, one rotation and one scale parameter, for particular locations based on the latitudes and longitudes of three or more common stations. Perhaps the best results derive from polynomial expressions developed for coordinate differences, expressed in Cartesian ($\Delta x, \Delta y, \Delta z$) or ellipsoidal coordinates ($\Delta f, \Delta \lambda, \Delta h$), using a 3-D Helmert transformation. However, besides requiring seven parameters (three shift, one scale, and three rotation components), this approach is at its best when ellipsoidal heights are available for all the points involved. Where adequate information is available, software packages such as the NGS programs LEFTI or NADCON can provide geodetic quality coordinates.

Even if a local transformation is modeled with these techniques, the resulting NAD27 positions might still be plagued with relatively low accuracy. The NAD83 adjustment of the national network is based on nearly 10 times the number of observations that supported the NAD27 system. This larger quantity of data, combined with the generally higher quality of the measurements at the foundation of NAD83, can have some rather unexpected results. For example, when NAD27 coordinates are transformed into the new system, the shift of individual stations may be quite different from what the regional trend indicates. In short, when using control from both NAD83 and NAD27 simultaneously on the same project, surveyors have come to expect difficulty.

In fact, the only truly reliable method of transformation is not to rely on coordinates at all, but to return to the original observations themselves. It is important to remember, for example, that geodetic latitude and longitude, as other coordinates, are specifically referenced to a given datum and are not derived from some sort of absolute framework. But the original measurements, incorporated into a properly designed least-squares adjustment, can provide most satisfactory results.

Densification and Improvement of NAD83

The inadequacies of NAD27 and even NAD83 positions in some regions are growing pains of a fundamentally changed relationship. In the past, relatively few engineers and surveyors were employed in geodetic work. Perhaps the greatest importance of the data from the various geodetic surveys was that they furnished precise points of reference to which the multitude of surveys of lower precision could then be tied. This arrangement was clearly illustrated by the design of state plane coordinates systems, devised to make the national control network accessible to surveyors without geodetic capability.

However, the situation has changed. The gulf between the precision of local surveys and national geodetic work is virtually closed by GPS, and that has changed the relationship between local surveyors in private practice and geodesists. For example, the significance of state plane coordinates as a bridge between the two groups has been drastically reduced. Today's surveyor has relatively easy and direct access to the geodetic coordinate systems themselves through GPS. In fact, the 1- to 2-ppm probable error in networks of relative GPS-derived positions frequently exceeds the accuracy of the first-order NAD83 positions intended to control them.

Fortunately, GPS surveyors have a chance to contribute to the solution of these difficulties. NGS will accept GPS survey data submitted in the correct format with proper supporting documentation. The process, known as *bluebooking*, requires strict adherence to NGS specifications. GPS measurements that can meet the criteria are processed, classified, and incorporated into the NGRS for the benefit of all GPS surveyors.

High Accuracy Reference Networks

Other significant work along this line is underway in the state-by-state supernet programs. The creation of *High Accuracy Reference Networks (HARN)* features cooperative ventures between NGS and the states, and often includes other organizations as well.

A station spacing of not more than about 60 miles and not less than about 15 miles is the objective in these statewide networks. The accuracy is intended to be 1 part-per-million, or better between stations. In other words, with heavy reliance on GPS observations, these networks are intended to provide extremely accurate, vehicle-accessible, regularly spaced control points with good overhead visibility. To ensure coherence, when the GPS measurements are complete, they are submitted to NGS for inclusion in a statewide readjustment of the existing NGRS covered by the state. Coordinate shifts of 0.3 to 1.0 m from NAD83 values have been typical in these readjustments.

The most important aspect of HARNs is the accuracy of their final positions. Entirely new orders of accuracy have been developed for GPS relative positioning techniques by the *Federal Geodetic Control Committee (FGCC)*. The FGCC is an organization chartered in 1968 and composed of representatives from 11 agencies of the federal government. It revises and updates surveying standards for geodetic control networks, among other duties. Its provisional standards and specifications for GPS work include Orders AA, A, and B, which are defined as having minimum geometric accuracies of 3 mm ±0.01 ppm, 5 mm ±0.1 ppm and 8 mm ±1 ppm, respectively, at the 95%, or 2σ, confidence level. The adjusted positions of HARN stations are designed to provide statewide coverage of at least B Order control as set out in these new standards.

The publication of up-to-date geodetic data, always one of the most important functions of NGS, is even more crucial today. The format of the data published by NGS has changed somewhat. NAD83 information includes, of course, the new ellipsoidal latitudes, longitudes, and azimuths. However, unlike the NAD27 data that provided the elevations of only some control points, NAD83 data includes elevations, or heights, for all marked stations.

Continuously Operating Reference Stations

About 1992, NGS began establishing a network of *Continuous Operating Reference Stations (CORS)* throughout the country. The original idea was to provide positioning for navigational and marine needs. There were about 50 CORS in

1996. Their positional accuracies are 3 cm horizontal and 5 cm vertical. They also must meet NOAA geodetic standards for installation, operation, and data distribution. In 1999 NGS brought an additional 33 Continuously Operating Reference Stations (CORS) on-line, resulting in a total of 165 stations in the National CORS Network.

The Continuously Operating Reference Stations in the NGS network are mostly to provide support for carrier phase observations. There are also many CORS available for differential correction of the code observations of DGPS. Tracking information is available for postprocessing on the internet.

NAD83 Positions and Plane Coordinates

The newly published data also include state plane coordinates in the appropriate zone. As before, the easting and northing are accompanied by the mapping angle and grid azimuths, but a scale factor is also included for easy conversions. *Universal Transverse Mercator (UTM)* coordinates are among the new elements offered by NGS in the published information for NAD83 stations.

These plane coordinates, both state plane and UTM, are far from an anachronism. The UTM projection has been adopted by the IUGG, the same organization that reached the international agreement to use GRS80 as the reference ellipsoid for the modern geocentric datum. NATO and other military and civilian organizations worldwide also use UTM coordinates for various mapping needs. UTM coordinates are often useful to those planning work that embraces large areas. In the United States, state plane systems based on the transverse Mercator projection, an oblique Mercator projection, and the Lambert conic map projection, grid every state, Puerto Rico, and the U.S. Virgin Islands into their own plane rectangular coordinate system. And GPS surveys performed for local projects and mapping are frequently reported in the plane coordinates of one of these systems.

For states with large east-west extent, the Lambert conic projection is used; this system uses a projection cone that is imagined to intersect the ellipsoid at standard parallels. When the cone is *developed;* that is, opened to make a plane, the ellipsoidal meridians become straight lines that converge at the cone's apex. The apex is also the center of the circular lines that represent the projections of the parallels of latitude.

Some states use both the Lambert conic and the transverse Mercator projections for individual zones within the state system (Figures 5.7 and 5.8). Some rely on the transverse Mercator projection alone. The transverse Mercator projection uses a projection cylinder whose axis is imagined to be parallel to the earth's equator and perpendicular to its axis of rotation. It intersects the ellipsoid along standard lines parallel to a central meridian. However, after the cylinder is developed all the projected meridians and parallels become curved lines.

Coordinates from these developed projections are given in reference to a Cartesian grid with two axes. Eastings are reckoned from an axis placed far west of the coordinate zone, adding a large constant value so all remain positive. Northings are reckoned from a line far to the south for the same reason.

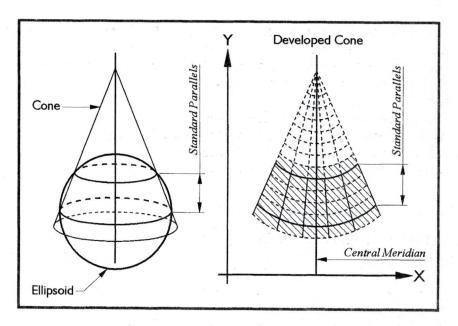

Figure 5.7. Lambert Conic Projection.

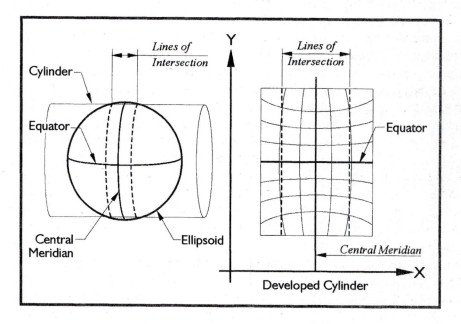

Figure 5.8. Transverse Mercator Projection.

The *x*-coordinate, the easting, and the *y*-coordinate, the northing, are expressed in either survey feet or international feet, depending on the state. NAD83 has required a redefinition of the state plane coordinate systems for the updated latitudes and longitudes. The constants now published by NGS are given in meters.

Both of these projections may be said to be *conformal*. Conformality means that an angle on the ellipsoid is preserved after mapping it onto the plane. This feature allows the shapes of small geographical features to look the same on the map as they do on the earth.

The UTM projection divides the world into 60 zones that begin at λ 180°, each with a width of 6° of longitude, extending from 84° Nϕ and 80° Sϕ. Its coverage is completed by the addition of two polar zones. The coterminous United States are within UTM zones 10 to 18.

The UTM grid is defined in meters. Each zone is projected onto a cylinder that is oriented in the same way as that used in the transverse Mercator state plane coordinates described above. The radius of the cylinder is chosen to keep the scale errors within acceptable limits. Coordinates of points from the reference ellipsoid within a particular zone are projected onto the UTM grid.

The intersection of each zone's central meridian with the equator defines its origin of coordinates. In the southern hemisphere, each origin is given the coordinates: easting = X_0 = 500,000 meters, and northing = Y_0 = 10,000,000 meters, to ensure that all points have positive coordinates. In the northern hemisphere, the values are: easting = X_0 = 500,000 meters, and northing = Y_0 = 0 meters, at the origin.

The scale factor grows from 0.9996 along the central meridian of a UTM zone to 1.00000 at 180,000 meters to the east and west. The state plane coordinate zones in the United States are limited to about 158 miles and so embrace a smaller range of scale factors than do the UTM zones. In state plane coordinates, the variance in scale is usually no more than 1 part in 10,000. In UTM coordinates the variance can be as large as 1 part in 2,500.

The distortion of positions attributable to the transformation of NAD83 geodetic coordinates into the plane grid coordinates of any one of these projections is generally less than a centimeter. Most GPS and land surveying software packages provide routines for automatic transformation of latitude and longitude to and from these mapping projections. Similar programs can also be purchased from the NGS. Therefore, for most applications of GPS, there ought to be no technical compunction about expressing the results in grid coordinates. However, given the long traditions of plane surveying, it can be easy for some to lose sight of the geodetic context of the entire process that produced the final product of a GPS survey presented in plane coordinates.

The Deflection of the Vertical

Other new elements in the information published by NGS for NAD83 positions include deflection of the vertical. The deflection of the vertical can be defined as the angle made by a line perpendicular to the geoid that passes through a point on the earth's surface with a line that passes through the same point but is perpendicular to the reference ellipsoid (Figure 5.9).

Described another way, the deflection of the vertical is the angle between the direction of a plumb line with the ellipsoidal normal through the same point. The

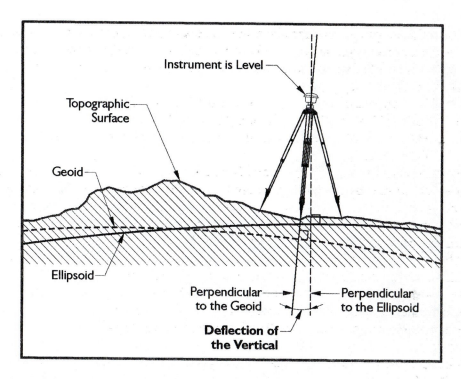

Figure 5.9. Deflection of the Vertical.

deflection of the vertical is usually broken down into two components, one in the plane of the meridian through the point and the other perpendicular to it. The first element is illustrated above.

When surveyors relied on astronomical observations for the determination of latitude, longitude, and azimuth, calculation of the deflection of the vertical was critical to deriving the corresponding ellipsoidal coordinates from the work. Today, if the orientation of a GPS vector is checked with an astronomic azimuth, some discrepancy should be expected. The deflection of the vertical is a result of the irregularity of the geoid and is mathematically related to separation between the geoid and the reference ellipsoid.

HEIGHTS

A point on the earth's surface is not completely defined by its latitude and longitude. In such a context there is, of course, a third element, that of height. Surveyors have traditionally referred to this component of a position as its elevation. One classical method of determining elevations is spirit leveling. As stated earlier, a level, correctly oriented at a point on the surface of the earth, defines a line parallel to the geoid at that point. Therefore, the elevations determined by level circuits are *orthometric;* that is, they are defined by their vertical distance above the geoid as it would be measured along a plumb line.

However, orthometric elevations are not directly available from the geocentric position vectors derived from GPS measurements. The vectors are not difficult to reduce to ellipsoidal latitude, longitude, and height because the reference ellipsoid is mathematically defined and clearly oriented to the earth if not perfectly geocentric. But the geoid defies such certain definition. As stated earlier, the geoid undulates with the uneven distribution of the mass of the earth and has all the irregularity that implies. In fact, the separation between the bumpy surface of the geoid and the smooth GRS80 ellipsoid worldwide varies from about + 85 meters to about –106 meters.

In the coterminous United States the variation is less. It is from about –8 meters to about –53 meters. As you can see, the geoid heights are negative because the geoid is beneath the ellipsoid. As a result, at any particular point the ellipsoidal height is actually smaller than the orthometric height in the conterminous United States.

Therefore, the only way a surveyor can convert an ellipsoidal height from a GPS observation on a particular station into a usable orthometric elevation is to know the extent of geoid-ellipsoid separation at that point.

Toward that end, major improvements have been made over the past quarter century or so in mapping the geoid on both national and global scales. This work has gone a long way toward the accurate determination of the geoid-ellipsoid separation, known as N. The formula for transforming ellipsoidal heights, h, into orthometric elevations, H, is (Figure 5.10):

$$H = h - N$$

Mean Sea Level

Orthometric heights, H, are directly related to local variations in gravity, of course. And GPS produces a much different kind of height, or elevation, that is not related to gravity at all. The GPS height is ellipsoidal, h, and also sometimes known as a geodetic height. The definition of the relationship between these requires knowledge of N, the geoidal height.

Geodetic leveling can provide an orthometric height, or elevation, directly. Unfortunately, it is sometimes incorrectly called an elevation above mean sea level. Actually mean sea level is the average height of the surface of the sea for all stages of the tide. It can also be defined as the arithmetic mean of elevation of the water's surface observed hourly over a 19-year cycle. These definitions do not describe a datum adequate for geodetic surveying since they vary with the time and place of measurement. And even though it was long considered an adequate approximation of the geoid, by these definitions mean sea level can vary considerably from the geoid. The definition of mean sea level that may be considered to eliminate this objection is, "The average location of the interface between ocean and atmosphere, over a period of time sufficiently long so that all the random and periodic variations of short duration average to zero." However, modern technol-

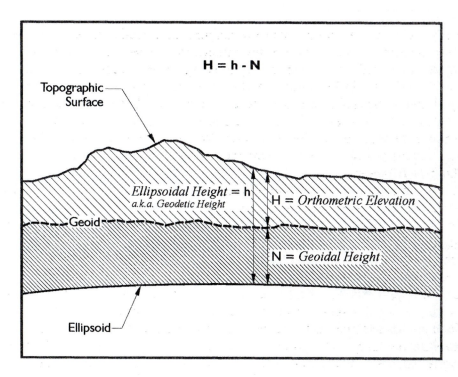

Figure 5.10. Formula for Transforming Ellipsoidal Heights.

ogy has improved dramatically the methods of measuring the actual geoid and the accuracy of geoidal models.

Because there are large complex variations in the geoid related to both the density and relief of the earth, geoid models, and interpolation software, have been developed to support the conversion of GPS elevations to orthometric elevations. For example, in early 1991 NGS presented a program known as GEOID90. This program allowed a user to find N, the geoidal height, in meters for any NAD83 latitude and longitude in the United States.

The GEOID90 model was computed at the end of 1990, using over a million gravity observations. It was followed by the GEOID93 model. It was computed at the beginning of 1993 using more than five times the number of gravity values used to create GEOID90. Both provided a grid of geoid height values in a 3 minutes of latitude by 3 minutes of longitude grid with an accuracy of about 10 cm. Next the GEOID96 model resulted in a gravimetric geoid height grid in a 2 minutes of latitude by 2 minutes of longitude grid.

Today, GEOID99 covers the coterminous United States, and it includes U.S. Virgin Islands, Puerto Rico, Hawaii, and Alaska. The grid is 1 degree of latitude by 1 degree of longitude and it is the first to combine gravity values with GPS ellipsoid heights on previously leveled benchmarks. According to NGS, "When comparing the GEOID99 model with GPS ellipsoid heights in the NAD 83 reference frame and leveling in the NAVD 88 datum, it is seen that GEOID99 has roughly a 4.6 cm absolute accuracy (one sigma) in the region of GPS on bench-mark coverage. In those states with sparse (150 km +) GPS on benchmark coverage, less point accuracy may be evident."

EXERCISES

1. What is the datum used for the GPS Navigation message?

   (a) NAD83
   (b) NAD27
   (c) GRS80
   (d) WGS84

2. What technology contributed the least number of positions to the least-squares adjustment of NAD83?
   (a) EDM baselines
   (b) TRANSIT Doppler positions
   (c) Conventional optical surveying
   (d) GPS

3. What is the highest minimum geometric accuracy specified by the FGCC standards for Order AA, Order A, and Order B GPS surveys at the 95%, or 2σ, confidence level?

   (a) 1 mm ± 0.01 ppm
   (b) 3 mm ± 0.01 ppm
   (c) 5 mm ± 0.1 ppm
   (d) 8 mm ± 1 ppm

4. In the State Plane Coordinate systems in the United States which mapping projection listed below is not used?

   (a) the transverse Mercator projection
   (b) the oblique Mercator projection
   (c) the Lambert conformal conic projection
   (d) the universal transverse Mercator projection

5. What information is necessary to convert an ellipsoidal height to an orthometric height?

   (a) the geoid height
   (b) the State Plane coordinate
   (c) the semimajor axis of the ellipsoid
   (d) the GPS Time

6. Which statement about the geoid is correct?

   (a)   The geoid's surface is always perpendicular to gravity.
   (b)   The geoid's surface is the same as mean sea level.
   (c)   The geoid's surface is always parallel with an ellipsoid.
   (d)   The geoid's surface is the same as the topographic surface.

7. What acronym is used to describe the cooperative ventures between NGS and the states to provide extremely accurate, vehicle-accessible, regularly spaced control points with good overhead visibility for GPS?

   (a)   ITRS
   (b)   HARN
   (c)   CORS
   (d)   NAVD

8. Which UTM zones cover the coterminous United States?

   (a)   Zones 10 North to 18 North
   (b)   Zones 1 North to 12 North
   (c)   Zones 6 North to 30 North
   (d)   Zones 20 North to 30 North

9. Which of the following organizations currently maintains the International Terrestrial Reference System, *ITRS?*

   (a)   NGS
   (b)   IERS
   (c)   BIH
   (d)   C&GS

10. Which of the following GEOID models combined GPS observations on existing benchmarks with gravity data?

   (a)   GEOID90
   (b)   GEOID96
   (c)   GEOID93
   (d)   GEOID99

ANSWERS and EXPLANATIONS

1. Answer is (d)

   Explanation: With very slight changes, GRS80 is the reference ellipsoid for the coordinate system known as the *World Geodetic System 1984 (WGS84).*

This datum has been used by the U.S. military since January 21, 1987 as the basis for the GPS Navigation message computations. Therefore, coordinates provided directly by GPS receivers are based in WGS84.

2. Answer is (d)

Explanation: It took more than 10 years to readjust and redefine the horizontal coordinate system of North America into what is now NAD83. More than 1.7 million positions derived from classical surveying techniques throughout the Western Hemisphere were involved in the least-squares adjustment. They were supplemented by approximately 30,000 EDM-measured baselines, 5,000 astronomic azimuths, and 650 Doppler stations positioned by the TRANSIT satellite system. Over 100 Very Long Baseline Interferometry (VLBI) vectors were also included. But GPS, in its infancy, contributed only five points

3. Answer is (b)

Explanation: The FGCC is an organization chartered in 1968 and composed of representatives from 11 agencies of the federal government. It revises and updates surveying standards for geodetic control networks, among other duties. Its provisional standards and specifications for GPS work include Orders AA, A, and B, which are defined as having minimum geometric accuracies of 3 mm ± 0.01 ppm, 5 mm ± 0.1 ppm, and 8 mm ± 1 ppm, respectively, at the 95 % or 2σ, confidence level.

4. Answer is (d)

Explanation: In the United States, state plane systems are based on the transverse Mercator projection, an oblique Mercator projection, and the Lambert conic map projection, grid. Every state, Puerto Rico, and the U.S. Virgin Islands has its own plane rectangular coordinate system.

5. Answer is (a)

Explanation: The geoid undulates with the uneven distribution of the mass of the earth and has all the irregularity that implies. In fact, the separation between the bumpy surface of the geoid and the smooth GRS80 ellipsoid varies from 0 up to ± 100 meters. Therefore, the only way a surveyor can convert an ellipsoidal height from a GPS observation on a particular station into a usable orthometric elevation is to know the extent of geoid-ellipsoid separation, also known as the *geoid height,* at that point.

Toward that end, major improvements have been made over the past quarter century or so in mapping the geoid on both national and global scales. This work has gone a long way toward the accurate determination of the geoid-ellipsoid separation, or geoid height, known as *N*. The formula for transforming ellipsoidal heights, *h*, into orthometric elevations, *H*, is

$$H = h - N$$

6. Answer is (a)

Explanation: The geoid is a representation of the Earth's gravity field. It is an equipotential surface that is everywhere perpendicular to the direction of gravity. In other words, it is perpendicular to a plumb line at every point.

Mean sea level is the average height of the surface of the sea for all stages of the tide. It was, and sometimes still is used as a reference for elevations. However, it is not the same as the geoid. Mean sea level departs from the surface of the geoid; these displacements are known as the sea surface topography. Neither is the ellipsoid, a smooth mathematically defined surface, always parallel to the bumpy geoid. Finally, the geoid is certainly not coincident with the topographic surface of the earth.

7. Answer is (b)

Explanation: High Accuracy Reference Networks (HARN) are cooperative ventures between NGS and the states, and often include other organizations as well.

With heavy reliance on GPS observations, these networks are intended to provide extremely accurate, vehicle-accessible, regularly spaced control points with good overhead visibility. To ensure coherence, when the GPS measurements are complete, they are submitted to NGS for inclusion in a statewide readjustment of the existing NGRS covered by the state. Coordinate shifts of 0.3 to 1.0 m from NAD83 values have been typical in these readjustments.

8. Answer is (a)

Explanation: The UTM projection divides the world into 60 zones that begin at λ 180°, each with a width of 6° of longitude, extending from 84° Nϕ and 80° Sϕ. Its coverage is completed by the addition of two polar zones. The coterminous United States are within UTM zones 10 to 18.

The UTM grid is defined in meters. Each zone is projected onto a cylinder that is oriented in the same way as that used in the transverse Mercator state plane coordinates described above. The radius of the cylinder is chosen to keep the scale errors within acceptable limits. Coordinates of points from the reference ellipsoid within a particular zone are projected onto the UTM grid.

The intersection of each zone's central meridian with the equator defines its origin of coordinates. In the southern hemisphere, each origin is given the coordinates: easting = X_0 = 500,000 meters, and northing = Y_0 = 10,000,000 meters, to ensure that all points have positive coordinates. In the northern hemisphere, the values are: easting = X_0 = 500,000 meters, and northing = Y_0 = 0 meters, at the origin. The scale factor grows from 0.9996 along the central meridian of a UTM zone to 1.00000 at 180,000 meters to the east and west.

9. Answer is (b)

Explanation: The best geocentric reference frame currently available is the International Terrestrial Reference Frame, *ITRF*. Its origin is at the center of mass of the whole earth including the oceans and atmosphere. The unit of length is the meter. The orientation of its axes was established as consistent with that of the IERS's predecessor, Bureau International de l'Heure, BIH, at the beginning of 1984.

Today, the ITRF is maintained by the International Earth Rotation Service, IERS, which monitors Earth Orientation Parameters, EOP for the scientific community through a global network of observing stations. This is done with GPS, Very Long Baseline Interferometry, *VLBI*, Lunar Laser Ranging, *LLR*, Satellite Laser Ranging, *SLR*, and Doppler Orbitography and Radiopositioning Integrated by Satellite (DORIS) and the positions of the observing stations are now considered to be accurate to the centimeter level.

The International Terrestrial Reference Frame (ITRF) is actually revised and published on a regular basis. And today NAD83 can be defined in terms of a best-fit transformation from ITRF96.

10. Answer is (d)

Explanation: Today, GEOID99 covers the coterminous United States, and it includes U.S. Virgin Islands, Puerto Rico, Hawaii, and Alaska. The grid is 1 degree of latitude by 1 degree of longitude and it is the first to combine gravity values with GPS ellipsoid heights on previously leveled benchmarks. According to NGS, "When comparing the GEOID99 model with GPS ellipsoid heights in the NAD 83 reference frame and leveling in the NAVD 88 datum, it is seen that GEOID99 has roughly a 4.6-cm absolute accuracy (one sigma) in the region of GPS on benchmark coverage. In those states with sparse (150 km +) GPS on benchmark coverage, less point accuracy may be evident."

6

Planning a Survey

ELEMENTS OF A GPS SURVEY DESIGN

If a GPS survey is carefully planned, it usually progresses smoothly. The technology has virtually conquered two stumbling blocks that have defeated the plans of conventional surveyors for generations. Inclement weather does not disrupt GPS observations, and a lack of intervisibility between stations is of no concern whatsoever, at least in static GPS. Still, GPS is far from so independent of conditions in the sky and on the ground that the process of designing a survey can now be reduced to points-per-day formulas, as some would like. Even with falling costs, the initial investment in GPS remains large by most surveyors standards. However, there is seldom anything more expensive in a GPS project than a surprise.

How Much Planning Is Required?

New Standards

The Federal Geodetic Control Committee (FGCC) has written provisional accuracy standards for GPS relative positioning techniques. The older standards of first, second, and third order are classified under the group C in the new scheme. In the past, the cost of achieving first-order accuracy was considered beyond the reach of most conventional surveyors. Besides, surveyors often said that such results were far in excess of their needs anyway. The burden of the equipment, techniques, and planning that is required to reach its 2σ relative error ratio of 1 part in 100,000 was something most surveyors were happy to leave to government agencies. But the FGCC's proposed new standards of B, A, and AA, are, respectively, 10, 100, and 1000 times more accurate than the old first-order. The attainment of these accuracies does not require corresponding 10-, 100-, and 1000-fold increases in equipment, training, personnel, or effort. They are now well within the reach of private GPS surveyors both economically and technically.

New Design Criteria

These upgrades in accuracy standards not only accommodate GPS, they also have cast survey design into a new light for many surveyors. Nevertheless, it is not correct to say that every job suddenly requires the highest achievable accuracy, nor is it correct to say that every GPS survey now demands an elaborate design. In some situations, a crew of two, or even one surveyor on-site may carry a GPS survey from start to finish with no more of a plan than minute-to-minute decisions can provide, even though the basis and the content of those decisions may be quite different from those made in a conventional survey.

In areas that are not heavily treed and are generally free of overhead obstructions, the now-lower C group of accuracy may be possible without a prior design of any significance. But while it is certainly unlikely that a survey of photocontrol or work on a cleared construction site would present overhead obstructions problems comparable with a control survey in the Rocky Mountains, even such open work may demand preliminary attention. The location of vertical and horizontal control, access across privately owned property or government installations, or, if kinematic GPS is involved, the routes between the stations frequently require reconnaissance. Still, there is an approach to kinematic and pseudokinematic GPS that tends to minimizes such concerns.

Radial GPS

Radial GPS surveying calls for one receiver, the base or foothold, to remain on a control station throughout the work while one or more other rover receivers move from point to point (Figure 6.1). The advantages of this arrangement in the stop-and-go kinematic or pseudokinematic modes include the large number of positions that can be established in a short amount of time with little or no planning. However, since all the baselines so established must originate at the base station, redundancy requires repeat occupations by the rovers with the same or different base control station. These successive occupations ought to be separated by at least a quarter of an hour and less than a full day so the satellite constellation can reach a significantly different configuration. Further, lacking a preliminary design of the survey, the area must be free of overhead obstructions and sources of multipath. Of course, if the project is to be done with RTK or real-time DGPS, overhead obstructions are much less of a hindrance. Project points that are simultaneously near one another but far from the control station must be directly connected with a baseline to maintain the integrity of the survey. Finally, if the base receiver loses lock and it goes unnoticed, it will completely defeat the radial survey.

Some of the difficulties of the radial survey approach can be overcome by adding more base control station receivers. More efficiency can be achieved by adding additional roving receivers. However, as the number of receivers rises, the logistics become more complicated, and a survey plan becomes necessary. In other words, in large GPS projects, in projects where overhead obstructions are

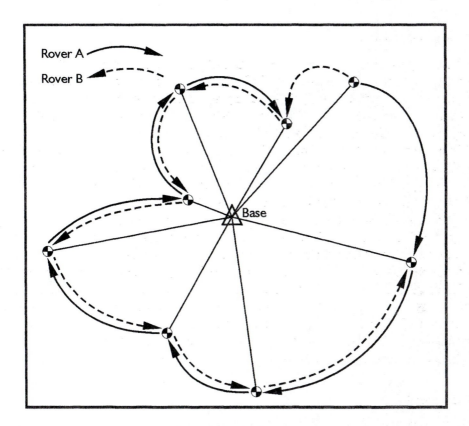

Figure 6.1. Radial GPS.

a factor, or where ties to the National Geodetic Reference System are important, some level of reconnaissance and design is required. If the survey crew is new to GPS or the higher orders of accuracy are called for, a survey design is certainly necessary.

Combining GPS Techniques

While RTK, DGPS kinematic, and pseudokinematic techniques are well suited to radial surveys for some topographic, photogrammetric, and GIS work, static and rapid-static are generally more appropriate for control surveys. The most efficient application of GPS to a particular project may very well be a combination of all these techniques, and each may require a different level of planning. Nevertheless, it is static and rapid-static GPS that require the fullest preliminary concern. The procedures presented here will be focused on the needs of that facet of GPS. While every element may not be necessary to a particular project or technique, it is very likely that some of them will.

Visiting the Site

Informing the Client

A visit to the site of a survey can often include a conference with the client, particularly when the client is new to the technology. Agreement on the cost and delivery date of a GPS project should be part of a larger body of information. GPS's reputation for extraordinary speed and accuracy can mislead some to expect miraculous, and impossible, results.

Even a small amount of understanding of the GPS surveying process goes a long way. For example, since the 3D Cartesian coordinates of GPS positions can be transformed into nearly any coherent coordinate system, a client may choose the form that is most useful to him or her. However, a mere list of coordinates may not be an adequate final product. The client may wish to have some formal account of the process that preceded the coordinates. A discussion of that report at the beginning of a project may save all concerned from subsequent misunderstandings.

The Lay of the Land

An initial visit to the site of the survey is not always possible, but it is almost always desirable. Although preliminary reconnaissance will certainly be cursory, general impressions may be formed on important questions that can be addressed more specifically later. For example, topography as it affects the line of sight between stations is of no concern on a GPS project, unless there are concerns about a radio link, but its influence on transportation from station to station is a primary consideration in designing the survey. Perhaps some areas are only accessible by helicopter or other special vehicle. Initial inquiries can be made. Roads may be excellent in one area of the project and poor in another. The general density of vegetation, buildings, or fences may open general questions of overhead obstruction or multipath. The pattern of land ownership, relative to the location of project points may raise or lower the level of concern about obtaining permission to cross property. Despite record information, the actual availability of horizontal and vertical control may be clearer on the ground of the project itself.

If night work, observations on or near highways, or unusual circumstances of any kind that may arouse public curiosity are anticipated, a visit with local law enforcement may be a good idea. Explanations at the earliest stages of a project can often eliminate difficulty later. The initial visit is also a good opportunity to learn about special safety regulations or other ordinances that will need to be satisfied during the project.

Client Participation

These and many more questions will arise and certain answers will probably be few. It is not uncommon that the client is more familiar with the area than the surveyor and can be a valuable resource in resolving many concerns. In fact, it is sometimes mutually convenient for the client to provide permissions, special vehicles, project point locations, or perform other tasks as part of the contractual arrangements.

Project Planning, Off-Site

Maps

Maps are particularly valuable resources for preparing a GPS survey design. Local government and private sources can sometimes provide the most appropriate mapping. Depending on the scope of the survey, various scales and type of maps can be useful. But the standard for planning GPS surveys remains the USGS 7.5- and 15-minute quad sheets.

A GPS survey plan usually begins in earnest with the plotting of all potential control and project points on a scale map of the area. USGS topographic maps from 1:24,000, the 7.5-minute quadrangle, to 1:62,500, the 15-minute quadrangle, to maps at 1:250,000 scale provide a good foundation for this preparation. The location of roads, vegetation, boundaries, power lines, landing areas, and a wide variety of pertinent information is available immediately from these maps.

Other mapping that may be helpful is available from various government agencies: for example, the U.S. Forest Service in the Department of Agriculture; the Department of Interior's Bureau of Land Management, Bureau of Reclamation, and National Park Service; the U.S. Fish and Wildlife Service in the Department of Commerce; and the Federal Highway Administration in the Department of Transportation are just a few of them. Even county and city maps should be considered, since they can sometimes provide the most timely information available. However, one vital element of the design is not available from any of these maps: the NGRS stations.

NGS Control Data Sheets

It is quite important to have the most up-to-date control information from NGS. A rectangular search based upon the range of latitudes and longitudes can now be performed on the NGS internet site. It is also possible to do a radial search, defining the region of the survey with one center position and a radius. Either search can display up to 100 *data sheets*. Depending on the capabilities of your browser, you may view them or save them to a file for further processing. You may also retrieve individual data sheets by the Permanent Identifier, *PID*, control point name, which is known as the *designation,* survey project identifier or USGS quad. It is best to ask for the desired horizontal and vertical information within a region that is somewhat larger than that which is contained by the boundaries of the survey. The internet address for NGS Data Sheets is http://www.ngs.noaa.gov/datasheet.html.

Here you will have several options. The standard DSDATA format is the traditional format of NGS data sheets. It is the format found on NGS paper products, tapes, diskettes, and CD-ROMs. There is a huge amount of information about survey monuments on each individual DSDATA record. They are sorted alphanumerically by station designation.

NGS also provides a map interface called *NGSmap* from which up to 32 data sheets may be retrieved.

The information available from an NGS control sheet is valuable at the earliest stage of a GPS survey (see Figure 6.2). In addition to the latitude and longitude, the published data include the state plane coordinates in the appropriate zones. The coordinates facilitate the plotting of the station's position on the project map.

The first line of each data sheet includes the retrieval date. Then the station's category is indicated. There are several, and among them are Continuously Operating Reference Station, Federal Base Network Control Station, and Cooperative Base Network Control Station.

This is followed by the station's designation, which is its name, and its Permanent Identifier, *PID*. Either of these may be used to search for the station in the NGS database. The PID is also found all along the left side of each data sheet record and is always 2 uppercase letters followed by four numbers.

The state, county, and USGS 7.5 minute quad name follow. Even though the station is located in the area covered by the quad sheet, it may not actually appear in the map.

Under the heading "Current Survey Control," you will find the latitude and longitude of the station in NAD83 and its height in NAVD88. Adjustment to NAD27 and NGVD29 datums are a thing of the past. However, these old values may be shown under *Superseded Survey Control.* Horizontal values may be either *Scaled,* if the station is a benchmark or *Adjusted,* if the station is indeed a horizontal control point.

When a date is shown in parentheses after NAD83 in the data sheet it means that the position has been readjusted since. Often these new adjustments are due to the station's inclusion in a State High Accuracy Reference Network, *HARN,* effort. There is more information on these cooperative projects in Chapter 5.

There are 13 sources of vertical control values shown on NGS data sheets. Here are a few of the categories. There is *Adjusted,* which are given to 3 decimal places and are derived from least-squares adjustment of precise leveling. Another category is *Posted,* which indicates that the station was adjusted after the general NAVD adjustment in 1991. When a station's elevation has been found by precise leveling but nonrigorous adjustment, it is called *Computed.*

Stations' vertical values are given to 1 decimal place if they are from GPS observation, *GPS Obs,* or vertical angle measurements, *Vert Ang.* And they have no decimal places if they were scaled from topographic map, *Scaled,* or found by conversion from NGVD29 values using the program known as VERTCON, *VERTCON.*

When they are available, earth-centered earth-fixed, *ECEF,* coordinates are shown. These are right-handed system, 3D Cartesian coordinates. They are the

The NGS Data Sheet

DATABASE = Sybase, PROGRAM = datasheet, VERSION = 6.30
Starting Datasheet Retrieval ....
1   National Geodetic Survey, Retrieval Date = December 17, 2000

```
KK1696 ********************************************************************************
KK1696  CBN              -  This is a Cooperative Base Network Control Station.
KK1696  DESIGNATION      -  JOG
KK1696  PID              -  KK1696
KK1696  STATE/COUNTY     -  CO / ARAPAHOE
KK1696  USGS QUAD        -  PARKER (1994)
KK1696
KK1696                         * CURRENT SURVEY CONTROL
KK1696
```

KK1696						
KK1696*	NAD 83 (1992)	-	39 34 05.17515 (N)	104 52 18.24505 (W)		ADJUSTED
KK1696*	NAVD 88	-	1796.3 (meters)	5893. (feet)		GPS OBS
KK1696						
KK1696	X	-	-1,263,970.470 (meters)			COMP
KK1696	Y	-	-4,759,798.648 (meters)			COMP
KK1696	Z	-	4,042,268.537 (meters)			COMP
KK1696	LAPLACE COOR	-	-5.62 (seconds)			DEFLEC99
KK1696	ELLIP HEIGHT	-	1779.26 (meters)			GPS OBS
KK1696	GEOID HEIGHT	-	-17.10 (meters)			GEOID99

```
KK1696
KK1696  HORZ ORDER    -  B
KK1696  ELLP ORDER    -  FOURTH      CLASS I
KK1696
KK1696. The horizontal coordinates were established by GPS observations and adjusted by
KK1696. the National Geodetic Survey in May 1992.
KK1696
KK1696. The orthometric height was determined by GPS observations and a high-resolution
KK1696. geoid model.
KK1696
KK1696. The X, Y, and Z were computed from the position and the ellipsoidal ht.
KK1696
KK1696. The Laplace correction was computed from DEFEC99 derived deflections.
KK1696
KK1696. The ellipsoidal height was determined by GPS observations and is referenced
KK1696. to NAD 83.
KK1696
KK1696. The geoid height was determined by GEOID99.
KK1696
```

KK1696;			North	East	Units	Scale	Converg.
KK1696;	SPC CO C	-	1,632,422.77	3,177,113.71	sFT	0.99996908	+0 23 46.5
KK1696;	SPC CO C	-	497,563.455	968,386.196	MT	0.99996908	+0 23 46.5
KK1696;	UTM 13	-	4,379,830.656	511,017.352	MT	0.99960149	+0 04 54.1

KK1696:			Primary Azimuth Mark		Grid Az
KK1696:	SPC CO C	-	JOG AZ MK		175 36 48.7
KK1696:	UTM 13	-	JOG AZ MK		175 55 41.1

```
KK1696
KK1696 |------------------------------------------------------------------------------|
```

KK1696 PID	Reference Object	Distance		Geod. Az
KK1696				dddmmss.s
KK1696 KK1695	DENVER INVERNESS TANK	66.890	METERS	11312
KK1696	JOG RM 1	13.428	METERS	13410
KK1696 KK1699	JOG AZ MK	APPROX. 0.7 KM		1760035.2
KK1696 KK1701	LITTLETON HONEYWELL CORP TANK	APPROX. 5.2 KM		2840833.9
KK1696	JOG RM 2	11.912	METERS	32655

```
KK1696 |------------------------------------------------------------------------------|
```

Figure 6.2. Format of NGS Data Sheet.

same type of X, Y, and Z coordinates presented in Chapter 5. These values are followed by the quantity which, when added to an astronomic azimuth, yields a geodetic azimuth; it is known as *the Laplace correction.*

It is important to note that NGS uses a clockwise rotation regarding the Laplace correction. The ellipsoid height per the NAD83 ellipsoid is shown, followed by the geoid height where the position is covered by NGS's GEOID program. Please see Chapter 5 for a more complete discussion of these values.

Survey Order and Class

Here the new accuracy standards mentioned earlier come into play. On NGS data sheets each adjusted control station will be assigned a horizontal, vertical (orthometric), and vertical (ellipsoid) order and class, where they apply.

Regarding horizontal control stations first-, second-, and third-order continue to be published under group C. However, these designations are now augmented by AA-, A-, and B-order stations as well. Horizontal AA-order stations have a relative accuracy of 3 mm ± 1:100,000,000 relative to other AA-order stations. Horizontal A-order stations have a relative accuracy of 5 mm ± 1:10,000,000 relative to other A-order stations. Horizontal B-order stations have a relative accuracy of 8 mm ± 1:1,000,000 relative to other A- and B-order stations.

Order and class continue to be published in first-, second-, and third-order for orthometric vertical control stations. Under the orders, class 0 is sometimes used. First-order, class 0 is used for a station whose tolerance is 2.0 mm or less. Second-order, class 0 is used for a station whose tolerance is 8.4 mm or less. Third-order, class 0 is used for a station whose tolerance is 12.0 mm or less. Posted benchmarks are given a distribution rate code from a to f, respectively, to indicate their reliability from 0 mm per km to 8 mm or more per km. Ellipsoid vertical control stations are also given order categories by NGS from first- to fifth- and each with a class 1 and 2, but the idea has not yet been adopted by the FGCC.

Coordinates

NGS data sheets also provide State Plane and UTM coordinates, the latter only for horizontal control stations. State Plane Coordinates are given in either U.S. Survey Feet or International Feet and UTM coordinates are given in meters. Azimuths to the primary azimuth mark are clockwise from north and scale factors for conversion from ellipsoidal distances to grid distances. This information may be followed by distances to reference objects. Coordinates are not given for azimuth marks or reference objects on the data sheet.

The Station Mark

Along with mark setting information, the type of monument and the history of mark recovery, the NGS data sheets provide a valuable *to-reach* description. It begins with the general location of the station. Then starting at a well-known location, the route is described with right and left turns, directions, road names, and the distance traveled along each leg in kilometers. When the mark is reached, the monument is described and horizontal and vertical ties are shown. Finally, there may be notes about obstructions to GPS visibility, etc.

Significance of the Information

The value of the description of the monument's location and the route used to reach it is directly proportional to the date it was prepared and the remoteness of its location. The conditions around older stations often change dramatically when the area has become accessible to the public. If the age and location of a station increases the probability that it has been disturbed or destroyed then reference monuments can be noted as alternatives worthy of on-site investigation. However, special care ought to be taken to ensure that the reference monuments are not confused with the station marks themselves.

Horizontal Control

At this stage, the choice of horizontal control amounts to finding candidates for actual reconnaissance. Excepting work tied to High Accuracy Reference Network (HARN) control, the so-called *supernet stations,* the accuracy of GPS measurements frequently exceed that of the stations used to control them. Since the final network of a GPS survey may well require constraint to NGS stations of somewhat inferior accuracy, those with the highest-available order should always be preferred. However, these provisional decisions about horizontal control require consideration of more than the published accuracy of the stations.

When geodetic surveying was more dependent on optics than electronic signals from space, horizontal control stations were set with station intervisibility in mind, not ease of access. Therefore it is not surprising that they are frequently difficult to reach. Not only are they found on the tops of buildings and mountains, they are also in woods, beside transmission towers, near fences, and generally obstructed from GPS signals. The geodetic surveyors that established them could hardly have foreseen a time when a clear view of the sky above their heads would be crucial to high-quality control.

In fact, it is only recently that most private surveyors have had any routine use for NGS stations. Many station marks have not been occupied for quite a long time. Since the primary monuments are often found deteriorated, overgrown, unstable, or destroyed, it is important that surveyors be well acquainted with the underground marks, R.M.'s and other methods used to perpetuate control stations.

Obviously, it is a good idea to propose reconnaissance of several more than the absolute minimum of three horizontal control stations. Fewer than three makes any check of their positions virtually impossible. Many more are usually required in a GPS route survey. In general, in GPS networks the more well-chosen horizontal control stations that are available, the better. Some stations will almost certainly prove unsuitable unless they have been used previously in GPS work or are part of a HARN.

Station Location

The location of the stations, relative to the GPS project itself, is also an important consideration in choosing horizontal control. For work other than route surveys, a handy rule of thumb is to divide the project into four quadrants and to choose at least one horizontal control station in each. The actual survey should have at least one horizontal control station in three of the four quadrants. Each of them ought to be as near as possible to the project boundary. Supplementary control in the interior of the network can then be used to add more stability to the network (Figure 6.3).

At a minimum, route surveys require horizontal control at the beginning, the end, and the middle. Long routes should be bridged with control on both sides of the line at appropriate intervals. The standard symbol for indicating horizontal control on the project map is a triangle.

Vertical Control

Those stations with a published accuracy high enough for consideration as vertical control are symbolized by an open square or circle on the map. Those stations that are sufficient for both horizontal and vertical control are particularly helpful and are designated by a combination of the triangle and square (or circle).

A minimum of four vertical control stations are needed to anchor a GPS network. A large project should have more. In general, the more high-order benchmarks that are available, the better. Vertical control is best located at the four corners of a project.

Orthometric elevations are best transferred by means of classic spirit leveling. When vertical control is too far removed from the project or when the benchmarks are obstructed, if project efficiency is not drastically impaired, such work should be built into the project plan. When the distances involved are too long, two independent GPS measurements may suffice to connect a benchmark to the project. However, it is important to recall the difference between the ellipsoidal heights available from a GPS observation and the orthometric elevations yielded by a level circuit. Further, third-order level work is not improved by beginning at a first-order benchmark. When spirit levels are planned to provide vertical control positions, special care may be necessary to ensure that the precision of the conventional work is as consistent as possible with the rest of the GPS survey (Figure 6.4).

Route surveys require vertical control at the beginning and the end. They should be bridged with benchmarks on both sides of the line at intervals from 5 to 10 km.

Plotting Project Points

A solid dot is the standard symbol used to indicate the position of project points. Some variation is used when a distinction must be drawn between those points that are in place and those that must be set. When its location is appropriate, it is always a good idea to have a vertical or horizontal control station serve double duty as a project point. While the precision of their plotting may vary, it is important that project points be located as precisely as possible, even at this preliminary stage.

First, the accuracy of the approximate coordinates of the project points later scaled from the map often depend on their original plotting. These approximate

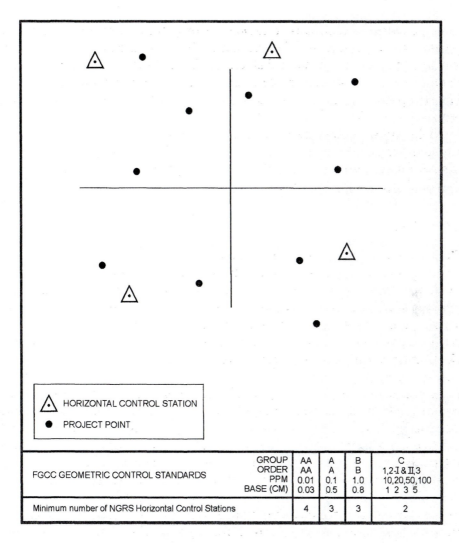

Figure 6.3. Horizontal Control and Project Points.

coordinates will give the GPS receiver, which eventually occupies the point, a reference position to begin its acquisition of satellites during the actual observation. Second, the subsequent observation schedule will depend to some degree on the arrangement of the baselines drawn on the map to connect the plotted points. Third, the preliminary evaluation of access, obstructions, and other information that can be derived from the map depends on the position of the project point relative to these features.

Evaluating Access

When all potential control and project positions have been plotted on the map and given a unique identifier, some aspects of the survey can be addressed a bit

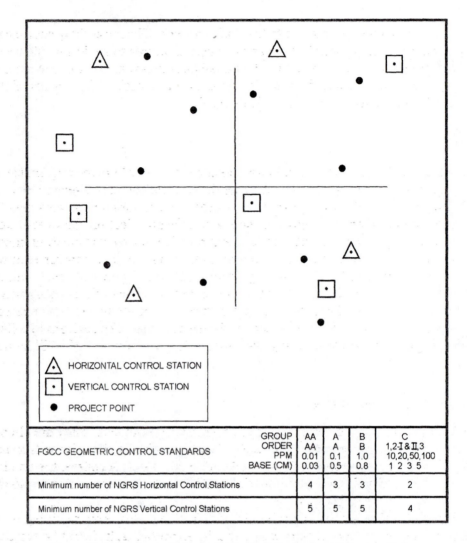

The following data accompanies the figure:

FGCC GEOMETRIC CONTROL STANDARDS	GROUP	AA	A	B	C
	ORDER	AA	A	B	1,2-I & II,3
	PPM	0.01	0.1	1.0	10,20,50,100
	BASE (CM)	0.03	0.5	0.8	1 2 3 5
Minimum number of NGRS Horizontal Control Stations		4	3	3	2
Minimum number of NGRS Vertical Control Stations		5	5	5	4

Legend:
- △ HORIZONTAL CONTROL STATION
- ⊡ VERTICAL CONTROL STATION
- ● PROJECT POINT

Figure 6.4. Horizontal Control, Vertical Control, and Project Points.

more specifically. If good roads are favorably located, if open areas are indicated around the stations, and if no station falls in an area where special permission will be required for its occupation, then the preliminary plan of the survey ought to be remarkably trouble-free. However, it is likely that one or more of these conditions will not be so fortunately arranged.

The speed and efficiency of transportation from station to station can be assessed to some degree from the project map. It is also wise to remember that while inclement weather does not disturb GPS observations whatsoever, without sufficient preparation it can play havoc with surveyors' ability to reach points over difficult roads or by aircraft.

In the case of a plan of survey for kinematic GPS, the route between the stations must be carefully examined on-site for any indication of overhead obstructions that may cause the receiver to lose lock en route. The most likely course can only be marked for reconnaissance at this stage. Features to avoid include trees, tunnels, bridges that cross over a road, and tall buildings that are near the road. A strategy where loss of lock is unavoidable is to set control on both sides of the obstruction so the receiver can be reinitialized.

Planning Offsets

If control stations or project points are located in areas where the map indicates that topography or vegetation will obstruct the satellite's signals, alternatives may be considered. A shift of the position of a project point into a clear area may be possible where the change does not have a significant effect on the overall network. A control station may also be the basis for a less obstructed position, transferred with a short level circuit or traverse. Of course, such a transfer requires availability of conventional surveying equipment on the project (which will be covered later). In situations where such movement is not possible, careful consideration of the actual paths of the satellites at the station itself during on-site reconnaissance may reveal enough windows in the gaps between obstructions to collect sufficient data by strictly defining the observation sessions (which will be discussed later, also).

Planning Azimuth Marks

Azimuth marks are a common requirement in GPS projects. They are almost always a necessary accompaniment to GPS stations when a client intends to use them to control subsequent conventional surveying work. Of course, the line between the station and the azimuth mark should be as long as convenience and the preservation of line-of-sight allows.

It is wise to take care that short baselines do not degrade the overall integrity of the project. Occupations of the station and its azimuth mark should be simultaneous for a direct measurement of the baseline between them. Both should also be tied to the larger network as independent stations. There should be two or more occupations of each station when the distance between them is less than 2 km.

While an alternative approach may be to derive the azimuth between a GPS station and its azimuth mark with an astronomic observation, it is important to remember that a small error, attributable to the deflection of the vertical, will be present in such an observation. The small angle between the plumb line and a normal to the ellipsoid at the station can either be ignored or removed with a Laplace correction.

Obtaining Permissions

Another aspect of access can be considered when the project map finally shows all the pertinent points. Nothing can bring a well-planned survey to a halt faster than a locked gate, an irate landowner, or a government official who is convinced he should have been consulted, previously. To the extent that it is possible from the available mapping, affected private landowners and government jurisdictions should be identified and contacted. Taking this precaution at the earliest stage of the survey planning can increase the chance that the sometimes long process of obtaining permissions, gate keys, badges, or other credentials has a better chance of completion before the survey begins.

Any aspect of a GPS survey plan derived from examining maps must be considered preliminary. Most features change with time, and even those that are relatively constant cannot be portrayed on a map with complete exactitude. Nevertheless, steps toward a coherent workable design can be taken using the information they provide.

Some GPS Survey Design Facts

Though much of the preliminary work in producing the plan of a GPS survey is a matter of estimation, some hard facts must be considered, too. For example, the number of GPS receivers available for the work and the number of satellites above the observer's horizon at a given time in a given place are two ingredients that can be determined with some certainty.

Software Assistance

Most GPS software packages provide users with routines that help them determine the satellite *windows,* the periods of time when the largest number of satellites are simultaneously available. Now that the GPS system is operational and a full constellation of satellites are in orbit, observers are virtually assured of 24-hour, four-satellite coverage. This assurance is a welcome relief from the forced downtime in the early days of GPS. The delays that were caused by periods when the satellites in view numbered three and fewer are now virtually eliminated. However, the mere presence of four satellites above an observer's horizon does not guarantee collection of sufficient data. Therefore, despite the virtual certainty that at least four satellites will be available, evaluation of their configuration as expressed in the position dilution of precision (PDOP) is still crucial in planning a GPS survey.

PDOP

In GPS, the receiver's position is derived from the simultaneous solution of vectors between it and at least four satellites. The quality of that solution depends, in large part, on the distribution of the vectors. For example, any position determined when the satellites are crowded together in one part of the sky will be unreliable, because all the vectors will have virtually the same direction. Given the ephemeris of each satellite, the approximate position of the receiver, and the time of the planned observation, a computer can predict such an unfavorable configuration and indicate the problem by giving the PDOP a large number. The GPS survey planner, on notice that the PDOP is large for a particular period of time, should consider an alternate observation plan.

On the other hand, when one satellite is directly above the receiver and three others are near the horizon and 120° in azimuth from one another, the arrangement is nearly ideal for a four-satellite constellation. The planner of the survey would be likely to consider such a window. However, more satellites would improve the resulting position even more, as long as they are well distributed in the sky above the receiver. In general, the more satellites, the better. For example, if the planner finds eight satellites will be above the horizon in the region where the work is to be done and the PDOP is below 2, that window would be a likely candidate for observation.

There are other important considerations. The satellites are constantly moving in relation to the receiver and to each other. Satellites rise and set and the PDOP is constantly changing. Within all this movement, the GPS survey designer must have some way of correlating the longest and most important baselines with the longest windows, the most satellites, and the lowest PDOP. Most GPS software packages, given a particular location and period of time, can provide illustrations of the satellite configuration.

Polar Plot

One such diagram is a plot of the satellite's tracks drawn on a graphical representation of the upper half of the celestial sphere with the observer's zenith at the center and perimeter circle as the horizon. The azimuths and elevations of the satellites above the specified mask angle are connected into arcs that represent the paths of all available satellites. The utility of this sort of drawing has lessened with the completion of the GPS constellation. In fact, there are so many satellites available that the picture can become quite crowded and difficult to decipher.

Another printout is a tabular list of the elevation and azimuth of each satellite at time intervals selected by the user.

An Example

The position of point Morant in the Table 6.1 needed expression to the nearest minute only, a sufficient approximation for the purpose. The ephemeris data were 5 days old when the chart was generated by the computer, but the data were still an adequate representation of the satellite's movements to use in planning. The mask angle was specified at 15°, so the program would consider a satellite set when it moved below that elevation angle. The zone time was Pacific Daylight Time, 7 hours behind Coordinated Universal Time, UTC. The full constellation provided 24 healthy satellites, and the sampling rate indicated that the azimuth and elevation of those above the mask angle would be shown every 10 minutes.

At 0:00 hour satellite PRN 2 could be found on an azimuth of 219° and an elevation of 16° above the horizon by an observer at 36°45′Nφ and 121°45′Wλ. The table indicates that PRN 2 was rising, and got continually higher in the sky

Table 6.1. Azimuth and Elevation Table

Satellites Azimuth and Elevation Table

Point: *Morant*
Date: *Wed., Sept. 29, 1993*
24 Satellites: *1 2 3 7 9 12 13 14 15 16 17 18 19 20 21 22 23 24 25 26 27 28 29 31*
Sampling Rate: *10 minutes*

Lat *36:45:0 N* Lon *121:45:0W*
Mask Angle: *15 (deg)*

Ephemeris: *9/24/93*
Zone: *Time Pacific Day (-7)*

Time	El Az	El Az	El Az	El Az	El Az	El Az	El Az	El Az	PDOP
SV	2	16	18	19	27	28	29	31	
				constellation of 8 SVs					
0:00	16 219	15 317	77 121	66 330	41 287	23 65	36 129	30 109	1.7
0:10	20 221	18 314	73 131	67 341	44 292	22 60	32 132	33 104	1.8
0:20	24 223	20 310	68 137	68 353	47 297	21 56	28 135	35 99	1.8
0:30	28 226	22 306	64 142	68 5	50 302	20 51	24 138	36 93	1.9
0:40	32 229	23 302	59 146	67 17	52 308	18 48	20 140	37 88	1.8
0:50	36 232	24 297	54 148	66 28	55 314	16 44	16 142	38 82	1.8
SV	2	16	18	19	27	31			
				constellation of 6 SVs					
1:00	40 235	24 293	49 151	65 39	58 320	38 76			3.0
1:10	43 239	24 288	44 153	63 49	61 328	37 70			3.0
1:20	47 244	24 283	40 155	61 57	64 336	36 64			2.8
1:30	51 249	23 278	35 156	59 65	66 345	34 60			2.6
SV	2	7	16	18	19	27	31		
				constellation of 7 SVs					
1:40	54 254	16 186	22 273	30 157	56 73	68 356	32 55		2.3
1:50	57 260	21 186	20 269	26 158	53 79	70 9	29 51		2.2
2:00	60 268	25 186	19 264	22 159	50 85	71 23	26 48		2.0
SV	2	7	16	18	19	26	27	31	
				constellation of 8 SVs					
2:10	66 276	30 185	16 260	17 160	47 91	15 319	71 38	23 45	1.7

for the 2 hours and 10 minutes covered by the chart. The satellite PRN 16 was also rising at 0:00 but reached its maximum altitude at about 1:10 and began to set. Unlike PRN 2, PRN 16 was not tabulated in the same row throughout the chart. It was supplanted when PRN 7 rose above the mask angle and PRN 16 shifted one column to the right. The same may be said of PRN 18 and PRN 19. Both of these satellites began high in the sky, unlike PRN 28 and PRN 29. They were just above 15° and setting when the table began and set after approximately 1 hour of availability. They would not have been seen again at this location for about 12 hours.

This chart indicated changes in the available constellation from eight space vehicles, *SVs,* between 0:00 and 0:50, six between 1:00 and 1:30, seven from 1:40 to 2:00, and back to eight at 2:10. The constellation never dipped below the minimum of four satellites, and the PDOP was good throughout. The PDOP varied between a low of 1.7 and a high of 3.0. Over the interval covered by the table, the PDOP never reached the unsatisfactory level of 5 or 6, which is when a planner should avoid observation.

Choosing the Window

Using this chart, the GPS survey designer might well have concluded that the best available window was the first. There was nearly an hour of eight-satellite data with a PDOP below 2. However, the data indicated that good observations could be made at any time covered here, except for one thing: it was the middle of the night. When a small number of satellites were available in the early days of GPS, the discomfort of such observations were ignored from necessity. With a full constellation, the loss of sleep can be avoided, and the designer may look at a more convenient time of day to begin the fieldwork.

Ionospheric Delay

It is worth noting that the ionospheric error is usually smaller after sundown. In fact, the FGCC specifies two-frequency receivers for daylight observations that hope to meet AA-, A-, and B-order accuracy standards, due, in part, to the increased ionospheric delay during those hours. There are provisions for compensation by modeling the error with two-frequency data from other sources where only single-frequency receivers are available. However, the specification illustrates the importance of considering atmospheric error sources.

An Example

Table 6.2, later in the day, covers a period of two hours when a constellation of five and six satellites was always available. However, through the first hour, from 6:30 to 7:30, the PDOP hovered around 5 and 6. For the first half of that hour, four of the satellites—PRN 9, PRN 12, PRN 13, and PRN 24—were all near the same elevation. During the same period, PRN 9 and PRN 12 were only approxi-

Table 6.2. Azimuth and Elevation Table Later in the Day

Satellites Azimuth and Elevation Table

Point: *Morant*

Date: *Wed., Sept. 29, 1993*

24 Satellites: *1 2 3 7 9 12, 13 14 15 16 17 18 19 20 21 22 23 24 25 26 27 28 29 31*

Sampling Rate: *10 minutes*

Lat *36:45:0 N* Lon *121:45:0W*

Mask Angle: *15 (deg)*

Ephemeris: *9/24/93*

Zone: *Time Pacific Day (-7)*

Time	El	Az	El	Az	El	Az	El	Az	El	Az	El	Az	PDOP
SV	7		9		12		13		24				
					constellation of 5 SVs								
6:30	28	54	60	271	61	319	62	15	48	177			6.3
6:40	24	57	60	261	66	314	57	19	53	176			6.0
6:50	21	60	59	252	70	305	53	22	58	175			5.3
7:00	18	62	57	243	73	292	49	25	63	172			4.6
SV	9		12		13		20		24				
					constellation of 5 SVs								
7:10	54	235	74	274	44	28	16	308	68	169			4.8
7:20	51	229	74	255	40	32	20	310	72	163			5.7
7:30	47	224	72	238	37	35	23	311	77	153			5.1
7:40	43	219	68	226	33	38	27	313	80	134			4.0
SV	9		12		13		16		20		24		
					constellation of 6 SVs								
7:50	39	215	64	218	29	41	16	149	31	314	81	102	2.1
8:00	35	212	59	213	26	45	19	146	36	314	80	73	2.3
8:10	31	209	54	209	23	48	23	143	40	315	76	57	2.4
8:20	27	207	49	206	19	52	27	140	44	314	72	49	2.5
8:30	23	204	44	204	16	55	30	137	48	314	67	45	2.5

mately 50° apart in azimuth, as well. Even though a sufficient constellation of satellites was constantly available, the survey designer may well have considered only the last 30 to 50 minutes of the time covered by this chart as suitable for observation.

There is one caution, however. Azimuth-elevation tables are a convenient tool in the division of the observing day into sessions, but it should not be taken for granted that every satellite listed is healthy and in service. For the actual availability of satellites and an update on atmospheric conditions, it is always wise to call the recorded message on the United States Coast Guard hotline at (703) 313-5907 or on-line you can check *GPS Status Message* at http://www.navcen.uscg.mil/ before and after a project. In the planning stage, the call can prevent creation of a design dependent on satellites that prove unavailable. Similarly, after the field-work is completed, it can prevent inclusion of unhealthy data in the postprocessing.

Supposing that the period from 7:40 to 8:30 was found to be a good window, the planner may have regarded it as a single 50-minute session, or divided it into shorter sessions. One aspect of that decision was probably the length of the baseline in question. In static GPS, a long line of 30 km may require 50 minutes of six-satellite data, but a short line of 3 km may not. If the planned survey was not done by static GPS, but instead with rapid-static, a 10-minute session may have been sufficient— in kinematic work the required session may be even shorter. Therefore, another aspect of the decision as to how the window was divided probably depended on the anticipated GPS surveying technique. A third consideration was probably the approximation of the time necessary to move from one station to another. More about the length of the baselines and estimated transportation times later.

Naming the Variables

The next step in the GPS survey design is drawing the preliminary plan of the baselines on the project map. Once some idea of the configuration of the baselines has been established, an observation schedule can be organized. Toward that end, the FGCC has developed a set of formulas provided in Appendix F of their provisional *Geometric Geodetic Accuracy Standards and Specifications for Using GPS Relative Positioning Techniques.* Those formulas will be used here.

For illustration, suppose that the project map (Figure 6.5) includes horizontal control, vertical control, and project points for a planned GPS network. They will be symbolized by m. There are four dual-frequency GPS receivers available for this project. They will be symbolized by r. There will be five observation sessions each day during the project. They will be symbolized by d. To summarize:

m = total number of stations (existing and new) = 14

d = number of possible observing sessions per observing day = 5

r = number of receivers = 4 dual frequency

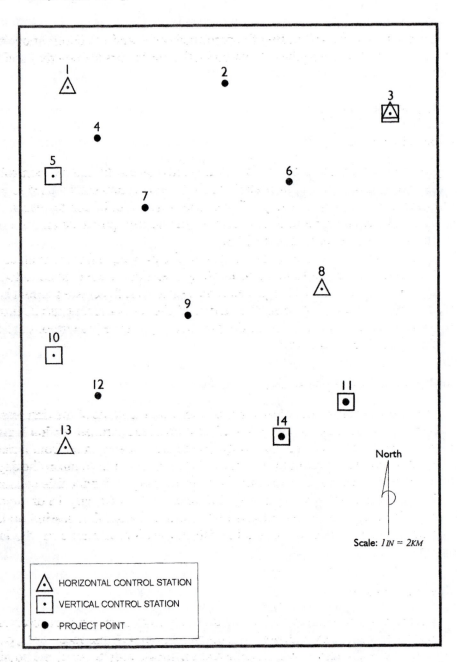

Figure 6.5. Project Map.

The design developed from this map must be preliminary. The session for each day of observation will depend on the success of the work the day before. Despite best efforts, any plan based on a map must be provisional until the baseline lengths, the obstructions at the observation sites, the transportation difficulties, the ionospheric disturbances, and the satellite geometry are actually known. Those questions can only be answered during the reconnaissance and the observations that follow. Even though these equivocations apply, the next step is to draw the baselines measurement plan.

Drawing the Baselines

Horizontal Control

A good rule of thumb is to verify the integrity of the horizontal control by observing baselines between these stations first. The vectors can be used to both corroborate the accuracy of the published coordinates and later to resolve the scale, shift, and rotation parameters between the control positions and the new network that will be determined by GPS.

These baselines are frequently the longest in the project, and there is an added benefit to measuring them first. By processing a portion of the data collected on the longest baselines early in the project, the degree that the sessions could have been shortened without degrading the quality of the measurement can be found. This test may allow improvement in the productivity on the job without erosion of the final positions (Figure 6.6).

Julian Day in Naming Sessions

The table at the bottom of Figure 6.6 indicates that the name of the first session connecting the horizontal control is 49-1. The date of the planned session is given in the Julian system. Taken most literally, Julian dates are counted from January 1, 4713 B.C. However, most practitioners of GPS use the term to mean the day of the current year measured consecutively from January 1. Under this construction, since there are 31 days in January, Julian day 49 is February 18 of the current year. The designation 49-1 means that this is to be the first session on that day. Some prefer to use letters to distinguish the session. In that case, the label would be 49-A.

Independent Lines

This project will be done with four receivers. The table shows that receiver A will occupy point 1; receiver B, point 3; receiver C, point 8; and receiver D, point 13 in the first session. However, the illustration shows only three of the possible six baselines that will be produced by this arrangement. Only the *independent*, also known as *nontrivial*, lines are shown on the map. The three lines that are not drawn are called *trivial*, and are also known as *dependent lines*. This idea is based on restricting the use of the lines created in each observing session to the absolute minimum needed to produce a unique solution.

Whenever four receivers are used, six lines are created. However, any three of those lines will fully define the position of each occupied station in relation to the others in the session. Therefore, the user can consider any three of the six lines

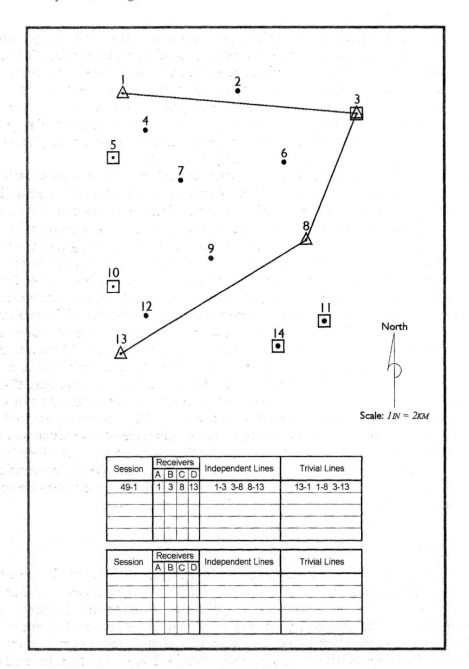

Figure 6.6. Drawing the Baselines.

independent. But once the decision is made, only those three baselines are included in the network. The remaining baselines are then considered trivial and discarded. In practice, the three shortest lines in a four-receiver session are almost always deemed the independent vectors, and the three longest lines are eliminated as trivial, or dependent. That is the case with the session illustrated.

Where r is the number of receivers, every session yields r-1 independent baselines. For example, four receivers used in 10 sessions would produce 30 independent baselines. It cannot be said that the shortest lines are always chosen to be the independent lines. Sometimes there are reasons to reject one of the shorter vectors due to incomplete data, cycle slips, multipath, or some other weakness in the measurements. Before such decisions can be made, each session will require analysis after the data has actually been collected. In the planning stage, it is best to consider the shortest vectors as the independent lines.

Another aspect of the distinction between independent and trivial lines involves the concept of error of closure, or loop closure. Loop closure is a procedure by which the internal consistency of a GPS network is discovered. A series of baseline vector components from more than one GPS session, forming a loop or closed figure, is added together. The closure error is the ratio of the length of the line representing the combined errors of all the vectors' components to the length of the perimeter of the figure. Any loop closures that only use baselines derived from a single common GPS session will yield an apparent error of zero, because they are derived from the same simultaneous observations. For example, all the baselines between the four receivers in session 49-1 of the illustrated project will be based on ranges to the same GPS satellites over the same period of time. Therefore, the trivial lines of 13-1, 1-8, and 3-13 will be derived from the same information used to determine the independent lines of 1-3, 3-8, and 8-13. It follows that, if the fourth line from station 13 to station 1 were included to close the figure of the illustrated session, the error of closure would be zero. The same may be said of the inclusion of any of the trivial lines. Their addition cannot add any redundancy or any geometric strength to the lines of the session, because they are all derived from the same data. If redundancy cannot be added to a GPS session by including any more than the minimum number of independent lines, how can the baselines be checked? Where does redundancy in GPS work come from?

Redundancy

If only two receivers were used to complete the illustrated project, there would be no trivial lines and it might seem there would be no redundancy at all. But to connect every station with its closest neighbor, each station would be have to be occupied at least twice, and each time during a different session. For example, with receiver A on station 1 and receiver B on station 2, the first session could establish the baseline between them. The second session could then be used to measure the baseline between station 1 and station 4. It would certainly be possible to simply move receiver B to station 4 and leave receiver A undisturbed on station 1. However, some redundancy could be added to the work if receiver A were reset. If it were recentered, replumbed, and its H.I. remeasured, some check on both of its occupations on station 1 would be possible when the network was completed. Under this scheme, a loop closure at the end of the project would have some meaning.

If one were to use such a scheme on the illustrated project and connect into one loop all of the 14 baselines determined by the 14 two-receiver sessions, the resulting error of closure would be useful. It could be used to detect blunders in the work, such as mis-measured H.I.s. Such a loop would include many different sessions. The ranges between the satellites and the receivers defining the baselines in such a circuit would be from different constellations at different times. On the other hand, if it were possible to occupy all 14 stations in the illustrated project with 14 different receivers simultaneously and do the entire survey in one session, a loop closure would be absolutely meaningless.

In the real world, such a project is not usually done with 14 receivers nor with 2 receivers, but with 3, 4, or 5. The achievement of redundancy takes a middle road. The number of independent occupations is still an important source of redundancy. In the two-receiver arrangement every line can be independent, but that is not the case when a project is done with any larger number of receivers. As soon as three or more receivers are considered, the discussion of redundant measurement must be restricted to independent baselines, excluding trivial lines.

Redundancy is then partly defined by the number of independent baselines that are measured more than once, as well as by the percentage of stations that are occupied more than once. While it is not possible to repeat a baseline without reoccupying its endpoints, it is possible to reoccupy a large percentage of the stations in a project without repeating a single baseline. These two aspects of redundancy in GPS—the repetition of independent baselines and the reoccupation of stations—are somewhat separate.

FGCC Standards for Redundancy

To meet order AA geometric accuracy standards, the FGCC requires three or more occupations on 80% of the stations in a project. Three or more occupations are necessary on 40, 20, and 10% of the stations for A, B, and C standards, respectively. When the distance between a station and its azimuth mark is less than 2 km, both points must be occupied at least twice to meet any standard above second order. All vertical control stations must be occupied at least twice for all orders of accuracy. Two or more occupations are required for all horizontal control station in order AA. The percentage requirements for repeat occupations on horizontal control stations drops to 75, 50, and 25% for A, B, and C, respectively. For new project points, reoccupation is mandated on 80, 50, and 10% of the stations in the project for A, B, and C, respectively.

The standards for repeat measurements of independent baselines in the FGCC provisional specifications note that an equal number of N-S and E-W vectors should be remeasured in a network. Of the independent baselines, 25% should be repeated in a project to meet order AA geometric accuracy standards. The standards require 15, 5, and 5% for orders A, B, and C, respectively.

Unless a project is to be *bluebooked;* that is, submitted to the NGS for inclusion in the national network, or there is a contractual obligation, there is usually no need to meet the letter of the specifications listed above. They are offered here as

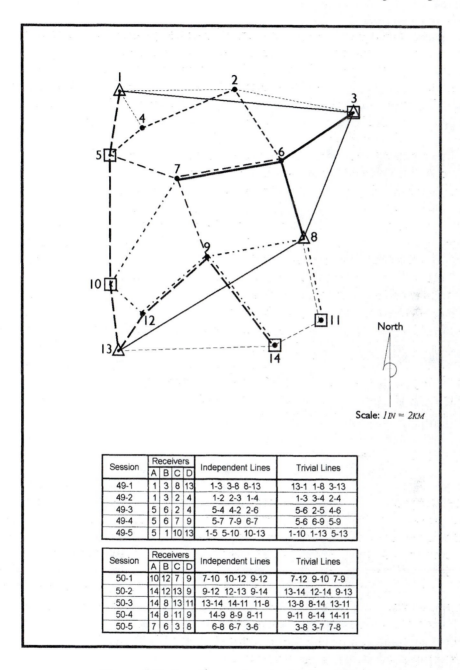

Session	Receivers				Independent Lines	Trivial Lines
	A	B	C	D		
49-1	1	3	8	13	1-3 3-8 8-13	13-1 1-8 3-13
49-2	1	3	2	4	1-2 2-3 1-4	1-3 3-4 2-4
49-3	5	6	2	4	5-4 4-2 2-6	5-6 2-5 4-6
49-4	5	6	7	9	5-7 7-9 6-7	5-6 6-9 5-9
49-5	5	1	10	13	1-5 5-10 10-13	1-10 1-13 5-13

Session	Receivers				Independent Lines	Trivial Lines
	A	B	C	D		
50-1	10	12	7	9	7-10 10-12 9-12	7-12 9-10 7-9
50-2	14	12	13	9	9-12 12-13 9-14	13-14 12-14 9-13
50-3	14	8	13	11	13-14 14-11 11-8	13-8 8-14 13-11
50-4	14	8	11	9	14-9 8-9 8-11	9-11 8-14 14-11
50-5	7	6	3	8	6-8 6-7 3-6	3-8 3-7 7-8

Figure 6.7. One Approach to Setting Baselines.

an indication of the level of redundancy that is necessary for high-accuracy GPS work.

Figure 6.7 shows one of the many possible approaches to setting up the baselines for this particular GPS project. The survey design calls for the horizontal control to be occupied in session 49-1. It is to be followed by measurements between two control stations and the nearest adjacent project points in session 49-2. As shown in the table at the bottom of Figure 6.7, there will be redundant occupations on stations 1 and 3. Even though the same receivers will occupy those points, their operators will be instructed to reset them at different H.I.s for the new session. A better, but probably less efficient, plan would be to occupy these stations with different receivers than were used in the first session.

Forming Loops

As the baselines are drawn on the project map for a static GPS survey, or any GPS work where accuracy is the primary consideration, the designer should remember that part of their effectiveness depends on the formation of complete geometric figures. When the project is completed, these independent vectors should be capable of formation into closed loops that incorporate baselines from two to four different sessions. In the illustrated baseline plan, no loop contains more than 10 vectors, no loop is more than 100 km long, and every observed baseline will have a place in a closed loop.

Finding the Number of Sessions

The illustrated survey design calls for 10 sessions, but the calculation does not include human error, equipment breakdown, and other unforeseeable difficulties. It would be impractical to presume a completely trouble-free project. The FGCC proposes the following formula for arriving at a more realistic estimate:

$$s = \frac{(m \cdot n)}{r} + \frac{(m \cdot n)(p-1)}{r} + k \cdot m$$

Where s =   the number of observing sessions,
         r =   the number of receivers,
         m =   the total number of stations involved.

But n, p, and k require a bit more explanation. The variable n is a representation of the level of redundancy that has been built into the network, based on the number of occupations on each station. The illustrated survey design includes more than two occupations on all but 4 of the 14 stations in the network. In fact, 10 of the 14 positions will be visited three or four times in the course of the survey. There are a total of 40 occupations by the 4 receivers in the 10 planned sessions. By dividing 40 occupations by 14 stations, it can be found that each station will be visited an average of 2.857 times. Therefore in the FGCC formula, the planned redundancy represented by factor n is equal to 2.857 in this project.

The experience of a firm is symbolized by the variable p in the formula. The division of the final number of actual sessions required to complete past projects by the initial estimation yields a ratio that can be used to improve future predictions. That ratio is the production factor, p. A typical production factor is 1.1.

A safety factor of 0.1, known as k, is recommended for GPS projects within 100 km of a company's home base. Beyond that radius, an increase to 0.2 is advised.

The substitution of the appropriate quantities for the illustrated project increases the prediction of the number of observation sessions required for its completion:

$$s = \frac{(mn)}{r} + \frac{(mn)(p-1)}{r} + km$$

$$s = \frac{(14)(2.857)}{4} + \frac{(14)(2.857)(1.1-1)}{4} + (0.2)(14)$$

$$s = \frac{40}{4} + \frac{4}{4} + 2.8$$

$$s = 10 + 1 + 2.8$$

Where s = 14 sessions (rounded to the nearest integer)

In other words, the 2-day, 10-session schedule is a minimum period for the baseline plan drawn on the project map. A more realistic estimate of the observation schedule includes 14 sessions. It is also important to keep in mind that the observation schedule does not include time for on-site reconnaissance.

Ties to the Vertical Control

The ties from the vertical control stations to the overall network are usually not handled by the same methods used with the horizontal control. The first session of the illustrated project was devoted to occupation of all the horizontal control stations. There is no similar method with the vertical control stations. First, the geoidal undulation would be indistinguishable from baseline measurement error. Second, the primary objective in vertical control is for each station to be adequately tied to its closest neighbor in the network.

If a benchmark can serve as a project point, it is nearly always advisable to use it, as was done with stations 11 and 14 in the illustrated project. A conventional level circuit can often be used to transfer a reliable orthometric elevation from vertical control station to a nearby project point.

Combining GPS Surveying Methods

A mixture of GPS surveying methods is frequently the most efficient approach to a particular project. Static surveying, such as that used in the illustrated project, or pseudokinematic GPS are both well suited to establishing the overall control for a project to set the stage for other methods. For example, they can be used to set control on either side of obstructions to prepare for receiver initialization in later kinematic observations.

EXERCISES

1. Which of the following information is not available from the NGS data sheet for a particular station?

   (a) the type of monument
   (b) the State Plane coordinates
   (c) the latitude and longitude of the primary azimuth mark
   (d) the Permanent Identifier

2. Which of the following FGCC geometric accuracy standards would be correctly included under C?

   (a) 5 mm $\pm$ 1:10,000,000
   (b) 1 cm $\pm$ 1:1,000,000
   (c) 8 mm $\pm$ 1:1,000,000
   (d) 3 mm $\pm$ 1:100,000,000

3. Which of the following is a quantity which, when added to an astronomic azimuth, yields a geodetic azimuth?

   (a) the primary azimuth
   (b) the mapping angle, also known as the convergence
   (c) the Laplace correction
   (d) the second term

4. How many nontrivial, or independent, and how many trivial, or dependent baselines are created in 1 GPS session using four receivers?

   (a) 1 independent baseline and 5 dependent baselines
   (b) 3 independent baselines and 1 dependent baseline
   (c) 2 independent baselines and 3 dependent baselines
   (d) 3 independent baselines and 3 dependent baselines

5. Which of the following statements concerning loop closure in a GPS network is not correct?

   (a) A loop closure that uses baselines from one GPS session will appear to have no error at all.
   (b) Loop closure is a procedure by which the internal consistency of a GPS network is discovered.

(c) No baselines should be excluded from a loop closure analysis.
(d) At the completion of a GPS control survey the independent vectors should be capable of formation into closed loops that incorporate baselines from two to four different sessions.

6. How many observation sessions will be required for a GPS control survey involving 20 stations that is to be done with 4 GPS receivers, a planned redundancy of 2, a production factor of 1.1, and a safety factor of 0.2?

   (a) 21 observation sessions
   (b) 17 observation sessions
   (c) 15 observation sessions
   (c) 12 observation sessions

7. Why is it advisable that successive occupations in a GPS survey be separated by at least a quarter of an hour and less than a full day when possible?

   (a) To eliminate multipath when it corrupted the first occupation of the station.
   (b) To allow the satellite constellation to reach a significantly different configuration than it had during the first occupation of the station.
   (c) To overcome an overhead obstruction during the first occupation of the station.
   (d) To eliminate receiver clock errors.

8. Which of the four following options describes a best configuration for the horizontal control of a GPS static control survey?

   (a) Divide the project into four quadrants and choose at least one horizontal control station near the project boundary in each quadrant.
   (b) Divide the project into three parts and choose at least one horizontal control station near the center of the project in each.
   (c) The primary base station should be located at the center of the project.
   (d) Draw a north-south line through the center of the project, and choose at least three horizontal control stations near that line.

9. Which of the following data sheet searches is not possible on the NGS internet site?

   (a) A rectangular search based upon the range of latitudes and longitudes.
   (b) A search using a particular station's geoid height.
   (c) A search using a particular stations Permanent Identifier, *PID*.
   (d) A radial search; that is, defining the region of the survey with one center position and a radius.

10. Which of the following is the standard symbol used in this book for project points in a GPS control survey?

   (a) A solid dot
   (b) A triangle
   (c) A square
   (d) A circle

ANSWERS AND EXPLANATIONS

1. Answer is (c)

   Explanation: NGS data sheets provide State Plane and UTM coordinates, the latter only for horizontal control stations. State Plane Coordinates are given in either U.S. Survey Feet or International Feet and UTM coordinates are given in meters. They also provide mark setting information, the type of monument, and the history of mark recovery. The data sheet certainly shows the station's designation, which is its name, and its Permanent Identifier, *PID*. Either of these may be used to search for the station in the NGS database. The PID is also found all along the left side of each data sheet record and is always two uppercase letters followed by four numbers. However, it does not show the latitude and longitude of azimuth marks. That information may sometimes be found on a data sheet devoted to the particular azimuth mark.

2. Answer is (b)

   Explanation: The geometric accuracy of 1 cm ± 1:1,000,000 is a familiar standard for first-order horizontal control. This designation and second- and third-order are now augmented by AA-, A-, and B- order stations as well. Horizontal AA-order stations have a relative accuracy of 3 mm ± 1:100,000,000 relative to other AA-order stations. Horizontal A-order stations have a relative accuracy of 5 mm ± 1:10,000,000 relative to other A-order stations. Horizontal B-order stations have a relative accuracy of 8 mm ± 1:1,000,000 relative to other A- and B-order stations.

3. Answer is (c)

   Explanation: The Laplace correction is a quantity which, when added to an astronomically observed azimuth, yields a geodetic azimuth. It is important to note that NGS uses a clockwise rotation regarding the Laplace correction. It can contribute several seconds of arc.

4. Answer is (c)

   Explanation: Where *r* is the number of receivers, every session yields *r*-1 independent baselines. Four receivers used in one session would produce 6 baselines. Of these 3 would be independent, or nontrivial baselines and 3 would be dependent, or trivial baselines. It cannot be said that the shortest lines are always chosen to be the independent lines. Sometimes there are reasons to reject one of the shorter vectors due to incomplete data, cycle slips, multipath, or some other weakness in the measurements. Before such decisions can be made, each session will require analysis after the data has actually been collected. In the planning stage, it is best to consider the shortest vectors as the independent lines, but once the decision is made, only those three baselines are included in the network.

5. Answer is (c)

   Explanation: Any loop closures that only use baselines derived from a single common GPS session will yield an apparent error of zero, because they are derived from the same simultaneous observations. When the project is completed, the independent vectors in the network, excluding the dependent, or trivial baselines, should be capable of formation into closed loops that incorporate baselines from two to four different sessions.

   In GPS the error of closure is valid for orders A and B when three or more independent baselines, from three or more GPS sessions are included in the loop. Order AA requires four independent observations be included. Loop closures for first-order and lower should include a minimum of two observing sessions.

6. Answer is (c)

   Explanation: The appropriate formula is

   $$s = \frac{(m \cdot n)}{r} + \frac{(m \cdot n)(p-1)}{r} + k \cdot m$$

   Where s = the number of observing sessions
   r = the number of receivers
   m = the total number of stations involved
   n = the planned redundancy
   p = the experience, or production factor
   k = the safety factor

   The variable n is a representation of the level of redundancy that has been built into the network, based on the number of occupations on each station. The experience of a firm is symbolized by the variable p in the formula. A typical production factor is 1.1. A safety factor of 0.1, known as k, is recommended for GPS projects within 100 km of a company's home base. Beyond that radius, an increase to 0.2 is advised.

The substitution of the appropriate quantities for the illustrated project increases the prediction of the number of observation sessions required for its completion:

$$s = \frac{(mn)}{r} + \frac{(mn)(p-1)}{r} + km$$

$$s = \frac{(20)(2)}{4} + \frac{(20)(2)(1.1-1)}{4} + (0.2)(20)$$

$$s = \frac{40}{4} + \frac{4}{4} + 4$$

$$s = 10 + 1 + 4$$

$$s = 15 \text{ sessions}$$

7. Answer is (b)

   Explanation: Successive occupations ought to be separated by at least a quarter of an hour and less than a full day so the satellite constellation can reach a significantly different configuration than that which it had during the first occupation. Please recall that GPS measurements are not actually made between the occupied stations, but directly to the satellites themselves.

8. Answer is (a)

   Explanation: For work other than route surveys, a handy rule of thumb is to divide the project into four quadrants and to choose at least one horizontal control station in each quadrant. The actual survey should have at least one horizontal control station in three of the four quadrants. Each of them ought to be as near as possible to the project boundary. Supplementary control in the interior of the network can then be used to add more stability to the network.

9. Answer is (b)

   Explanation: It is quite important to have the most up-to-date control information from NGS. A rectangular search based upon the range of latitudes and longitudes can now be performed on the NGS internet site. It is also possible to do a radial search, defining the region of the survey with one center position and a radius. You may also retrieve individual data sheets by the

Permanent Identifier, *PID,* control point name, which is known as the *designation,* survey project identifier, or USGS quad. However, you cannot retrieve a data sheet for a station based only on its geoid height.

10. Answer is (a)

Explanation: A solid dot is the standard symbol used to indicate the position of project points. Some variation is used when a distinction must be drawn between those points that are in place and those that must be set. When its location is appropriate, it is always a good idea to have a vertical or horizontal control station serve double duty as a project point. While the precision of their plotting may vary, it is important that project points be located as precisely as possible, even at this preliminary stage. A horizontal control point is shown as a triangle, and a vertical control point is shown as a square.

7

Observing

PREPARING TO OBSERVE

In the following chapter the emphasis is on the use of static GPS in control surveying.

The prospects for the success of a GPS project are directly proportional to the quality and training of the people doing it. The handling of the equipment, the on-site reconnaissance, the creation of field logs, and the inevitable last-minute adjustments to the survey design all depend on the training of the personnel involved for their success. There are those who say the operation of GPS receivers no longer requires highly qualified survey personnel. That might be true if effective GPS surveying needed only the pushing of the appropriate buttons at the appropriate time. In fact, when all goes as planned, it may appear to the uninitiated that GPS has made experienced field surveyors obsolete. But when the unavoidable breakdowns in planning or equipment occur, the capable people, who seemed so superfluous moments before, suddenly become indispensable.

Training

One of the great drawbacks of the education demanded by high technology work of all kinds is the resulting nonbillable time. The often quoted principle that 85 % of every employee's time ought to be billable is certainly violated during any period of extensive training. GPS presents ever-increasing efficiency and productivity to a surveying operation. But the rate of that growth and change cannot be used to advantage without an equally constant growth in the knowledge of the personnel charged with actually doing the work.

Equipment

Conventional Equipment

Most GPS projects require conventional surveying equipment for spirit-leveling circuits, offsetting horizontal control stations, and monumenting project points, among other things. It is perhaps a bit ironic that this most advanced surveying method also frequently has need of the most basic equipment. The use of brush hooks, machetes, axes, etc., can sometimes salvage an otherwise unusable position by removing overhead obstacles. Another strategy for overcoming such hindrances has been developed using various types of survey masts to elevate a separate GPS antenna above the obstructing canopy.

Flagging, paint, and the various techniques of marking that surveyors have developed over the years are still a necessity in GPS work. The pressure of working in unfamiliar terrain is often combined with urgency. Even though there is usually not a moment to spare in moving from station to station, a GPS surveyor frequently does not have the benefit of having visited the particular points before. In such situations, the clear marking of both the route and the station during reconnaissance is vital. Marking the route between stations in kinematic GPS carries the added importance of preventing the loss of the continuous satellite lock essential to the survey's success.

Despite the best route marking, a surveyor may not be able to reach the planned station, or, having arrived, finds some new obstacle or unanticipated problem that can only be solved by marking and occupying an impromptu offset position for a session. A hammer, nails, shiners, paint, etc. are essential in such situations.

In short, the full range of conventional surveying equipment and expertise have a place in GPS. For some, their role may be more abbreviated than it was formerly, but one element that can never be outdated is good judgment.

Safety Equipment

The high-visibility vests, cones, lights, flagmen, and signs needed for traffic control cannot be neglected in GPS work. Unlike conventional surveying operations, GPS observations are not deterred by harsh weather. Occupying a control station in a highway is dangerous enough under the best of conditions, but in the midst of a rainstorm, fog, or blizzard, it can be absolute folly without the proper precautions. And any time and trouble taken to avoid infraction of the local regulations regarding traffic management will be compensated by an uninterrupted observation schedule.

Weather conditions also affect travel between the stations of the survey, both in vehicles and on foot. Equipment and plans to deal with emergencies should be part of any GPS project. First aid kits, fire extinguishers, and the usual safety equipment are necessary. Training in safety procedures can be an extraordinary benefit, but perhaps the most important capability in an emergency is communication.

Communications

Whether the equipment is handheld or vehicle mounted, two-way radios or cell phones are used in most GPS operations. A link between surveyors can increase the efficiency and safety of a GPS project, but it is particularly valuable when last-minute changes in the observation schedule are necessary. When an observer is unable to reach a station or a receiver suddenly becomes inoperable, unless adjustments to the schedule can be made quickly, each end of all of the lines into the missed station will require reobservation.

Unless a receiver is collecting data continuously at a foothold or master station, communication is particularly important in pseudokinematic surveys. The success of both pseudokinematic and all types of static GPS hinges on all receivers collecting their data simultaneously. However, it is difficult to ensure reliable communication between receiver operators in geodetic surveys, especially as their lines grow longer.

One alternative to contact between surveyors is reliance on the preprogramming feature available on most GPS receivers today. This attribute usually allows the start-stop time, sampling rate, bandwidth, satellites to track, mask angle, data file name, and start position to be preset so that the operator need only set up the receiver at the appropriate station before the session begins. The receiver is expected to handle the rest automatically. In theory, this approach eliminates the chance for an operator error ruining an observation session by missing the time to power-up or improperly entering the other information. Theory falls short of practice here, and even if the procedure could eliminate those mistakes, entire categories of other errors remain unaddressed. Some advocate actually leaving receivers unattended in static GPS. This idea seems unwise on the face of it.

High-wattage, private-line FM radios are quite useful when line of sight is available between them or when a repeater is available. The use of cell phones may eliminate the communication problem in some areas, but probably not in remote locations. Despite the limitations of the systems available at the moment, achievement of the best possible communication between surveyors on a GPS project pays dividends in the long run.

GPS Equipment

Most GPS receivers capable of geodetic accuracy are designed to be mounted on a tripod, usually with a tribrach and adaptor. However, there is a trend toward bipod- or range-pole-mounted antennas. Some of these arrangements are designed for easy vehicle mounting to facilitate the quick setups needed in kinematic or pseudokinematic applications. Another advantage of these devices is that they ensure a constant height of the antenna above the station. The mismeasured height of the antenna above the mark is probably the most pervasive and frequent blunder in GPS control surveying.

The tape or rod used to measure the height of the antenna is sometimes built into the receiver, and sometimes a separate device. It is important that the H.I. be measured accurately and consistently in both feet and meters, without merely converting from one to the other mathematically. It is also important that the value be recorded in the field notes and, where possible, also entered into the receiver itself.

Where tribrachs are used to mount the antenna, the tribrach's optical centering should be checked and calibrated. It is critical that the effort to perform GPS surveys to an accuracy of centimeters not be frustrated by inaccurate centering or H.I. measurement. Since many systems measure the height of the antenna to the edge of the ground plane or to the exterior of the receiver itself, the calibration of the tribrach affects both the centering and the H.I. measurement. The resetting of a receiver that occupies the same station in consecutive sessions is an important source of redundancy for many kinds of GPS networks. However, integrity can only be added if the tribrach has been accurately calibrated.

The checking of the carrier phase receivers themselves is also critical to the control of errors in a GPS survey, especially when different receivers or different models of antennas are to be used on the same work. The zero baseline test is a method that may be used to fulfill equipment calibration specifications where a three-dimensional test network of sufficient accuracy is not available. As a matter of fact, the simplicity of this test is a big advantage. It is not dependent on special software or test network. This test can also be used to separate receiver difficulties from antenna errors.

Two or more receivers are connected to one antenna with a *signal,* or *antenna, splitter.* The antenna splitter can be purchased from specialty electronics shops and is also available on-line. An observation is done with the divided signal from the single antenna reaching both receivers simultaneously. Since the receivers are sharing the same antenna, satellite clock biases, ephemeris errors, atmospheric biases, and multipath are all canceled. The only remaining errors are attributable to random noise and receiver biases. The success of this test depends on the signal from one antenna reaching both receivers, but the current from only one receiver can be allowed to power the antenna. This test checks not only the precision of the receiver measurements, but also the processing software. The results of the test should show a baseline of only a few millimeters. In fact, the zero baseline is used in product testing by some GPS manufacturers.

Information is also available on NGS calibration baselines throughout the United States. You can learn more by visiting the NGS site at, http://www.ngs.noaa.gov/

Auxiliary Equipment

Tools to repair the ends of connecting cables, a simple pencil eraser to clean the contacts of circuit boards, or any of a number of small implements have saved more than one GPS observation session from failure. Experience has shown that GPS surveying requires at least as much resourcefulness, if not more, than conventional surveying.

The health of the batteries are a constant concern in GPS. There is simply nothing to be done when a receiver's battery is drained but to resume power as soon as possible. A backup power source is essential. Cables to connect a vehicle battery, an extra fully-charged battery unit, or both should be immediately available to every receiver operator.

Papers

The papers every GPS observer carries throughout a project ought to include emergency phone numbers; the names, addresses, and phone numbers of relevant property owners; and the combinations to necessary locks. Each member of the team should also have a copy of the project map, any other maps that are needed to clarify position or access, and, perhaps most important of all, the updated observation schedule.

The observation schedule for static GPS work will be revised daily based upon actual production (Table 7.1). It should specify the start-stop times and station for all the personnel during each session of the upcoming day. In this way, the schedule will not only serve to inform every receiver operator of his or her own expected occupations, but those of every other member of the project as well. This knowledge is most useful when a sudden revision requires observers to meet or replace one another.

RECONNAISSANCE

Station Data Sheet

The principles of good field notes have a long tradition in land surveying, and they will continue to have validity for some time to come. In GPS, the ensuing paper trail will not only fill subsequent archives, it has immediate utility. For example, the station data sheet is often an important bridge between on-site reconnaissance and the actual occupation of a monument.

Though every organization develops its own unique system of handling its field records, most have some form of the station data sheet. The document illustrated in Figure 7.1 is merely one possible arrangement of the information needed to recover the station.

The station data sheet can be prepared at any period of the project, but perhaps the most usual times are during the reconnaissance of existing control or immediately after the monumentation of a new project point. Neatness and clarity, always paramount virtues of good field notes, are of particular interest when the station data sheet is to be later included in the final report to the client. The overriding principle in drafting a station data sheet is to guide succeeding visitors to the station without ambiguity. A GPS surveyor on the way to observe the position for the first time may be the initial user of a station data sheet. A poorly written document could void an entire session if the observer is unable to locate the monument. A client, later struggling to find a particular monument with an inadequate data sheet, may ultimately question the value of more than the field notes.

Station Name

The station name fills the first blank on the illustrated data sheet. Two names for a single monument is far from unusual. In this case the vertical control station, officially named S 198, is also serving as a project point, number 14. But two

Table 7.1. Observation Schedule-Day 49

	Session 1 Start 7:10 Stop 8:10	8:10 to 8:40	Session 2 Start 8:40 Stop 9:50	9:50 to 10:15	Session 3 Start 10:15 Stop 11:15	11:15 to 11:30	Session 4 Start 11:30 Stop 12:30	12:30 to 14:00	Session 5 Start 14:00 Stop 15:00
Svs PRNs	9,12,13,16, 20, 24		3,12,13,16, 20,2 4		3,12,13,16,17, 20, 24		3,16,17, 20, 22, 23, 26		1,3,17, 21, 23, 26, 28
Receiver A Dan H.	Station 1 NGS Horiz. Control	Reset	Station 1 NGS Horiz. Control	Move	Station 5 NGS Benchmark	Reset	Station 5 NGS Benchmark	Reset	Station 5 NGS Benchmark
Receiver B Scott G.	Station 3 NGS V&H Control	Reset	Station 3 NGS V&H Control	Move	Station 6 Project Point	Reset	Station 6 Project Point	Move	Station 1 NGS Horiz. Control
Receiver C Dewey A.	Station 8 NGS Horiz. Control	Move	Station 2 Project Point	Reset	Station 2 Project Point	Move	Station 7 Project Point	Move	Station 10 NGS Benchmark
Receiver D Cindy E.	Station 13 NGS Horiz. Control	Move	Station 4 Project Point	Reset	Station 4 Project Point	Move	Station 9 Project Point	Move	Station 13 NGS Horiz. Control

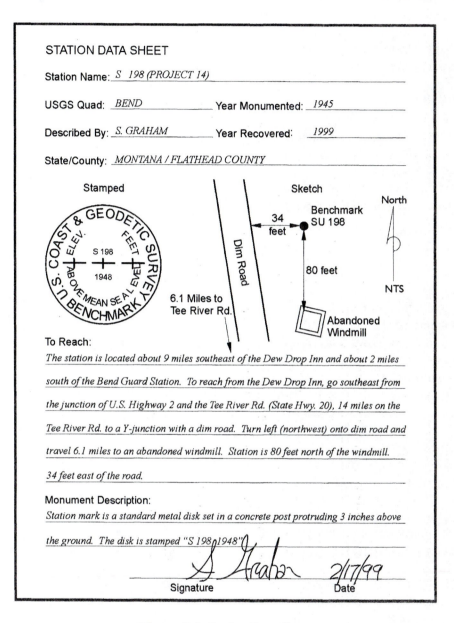

Figure 7.1. Station Data Sheet.

names purporting to represent the same position can present a difficulty. For example, when a horizontal control station is remonumented a number 2 is sometimes added to the original name of the station and it can be confusing. For example, it can be easy to mistake station, "Thornton 2" with an original station named "Thornton" that no longer exists. Both stations may still have a place in the published record, but with slightly different coordinates. Another unfortunate misunderstanding can occur when inexperienced field personnel mistake a reference mark, R.M., for the actual station itself. The taking of rubbings and/or close-up photographs is widely recommended to avoid such blunders regarding stations' names or authority.

Rubbings

The illustrated station data sheet provides an area to accommodate a rubbing. With the paper held on top of the monument's disk, a pencil is run over it in a zig-zag pattern producing a positive image of the stamping. This method is a bit more awkward than simply copying the information from the disk onto the data sheet, but it does have the advantage of ensuring the station was actually visited and that the stamping was faithfully recorded. Such rubbings or close-up photographs are required by the provisional FGCC *Geometric Geodetic Accuracy Standards and Specifications for Using GPS Relative Positioning Techniques* for all orders of GPS surveys.

Photographs

The use of photographs is growing as a help for the perpetuation of monuments. It can be convenient to photograph the area around the mark as well as the monument itself. These exposures can be correlated with a sketch of the area. Such a sketch can show the spot where the photographer stood and the directions toward which the pictures were taken. The photographs can then provide valuable information in locating monuments, even if they are later obscured. Still, the traditional ties to prominent features in the area around the mark are the primary agent of their recovery.

Quad Sheet Name

Providing the name of the appropriate state, county, and USGS quad sheet helps to correlate the station data sheet with the project map. The year the mark was monumented, the monument description, the station name, and the "to-reach" description all help to associate the information with the correct official control data sheet and, most importantly, the correct station coordinates.

To-Reach Descriptions

The description of the route to the station is one of the most critical documents written during the reconnaissance. Even though it is difficult to prepare the information in unfamiliar territory and although every situation is somewhat different, there are some guidelines to be followed. It is best to begin with the general location of the station with respect to easily found local features.

The description in Figure 7.1 relies on a road junction, a guard station, and a local business. After defining the general location of the monument, the description should recount directions for reaching the station. Starting from a prominent location, the directions should adequately describe the roads and junctions. Where the route is difficult or confusing, the reconnaissance team should not only describe the junctions and turns needed to reach a station, it is wise to also mark them with lath and flagging when possible. It is also a good idea to note gates. Even if they are open during reconnaissance, they may be locked later. When turns are called for, it is best to describe not only the direction of the turn, but the new course too. For example, in the description in Figure 7.1 the turn onto the dim road from the Tee River Road is described to the "left (northwest)." Roads and highways should carry both local names and designations found on standard highway maps. For example, in Figure 7.1 Tee River Road is also described as State Highway 20.

The "to-reach" description should certainly state the mileages as well as the travel times where they are appropriate, particularly where packing-in is required. Land ownership, especially if the owner's consent is required for access, should be mentioned. The reconnaissance party should obtain the permission to enter private property and should inform the GPS observer of any conditions of that entry. Alternate routes should be described where they may become necessary. It is also best to make special mention of any route that is likely to be difficult in inclement weather.

Where helicopter access is anticipated, information about the duration of flights from point to point, the distance of landing sites from the station, and flight time to fuel supplies should be included on the station data sheet.

Flagging and Describing the Monument

Flagging the station during reconnaissance may help the observer find the mark more quickly. On the station data sheet, the detailed description of the location of the station with respect to roads, fence lines, buildings, trees, and any other conspicuous features should include measured distances and directions. A clear description of the monument itself is important. It is wise to also show and describe any nearby marks, such as R.M.s, that may be mistaken for the station or aid in its recovery. The name of the preparer, a signature, and the date round out the initial documentation of a GPS station.

Visibility Diagrams

Obstructions above the mask angle of a GPS receiver must be taken into account in finalizing the observation schedule. A station that is blocked to some degree is not necessarily unusable, but its inclusion in any particular session is probably contingent on the position of the specific satellites involved.

An Example

The diagram in Figure 7.2 is widely used to record such obstructions during reconnaissance. It is known as a *station visibility diagram*, a *polar plot* or a *skyplot*. The concentric circles are meant to indicate 10° increments along the upper half

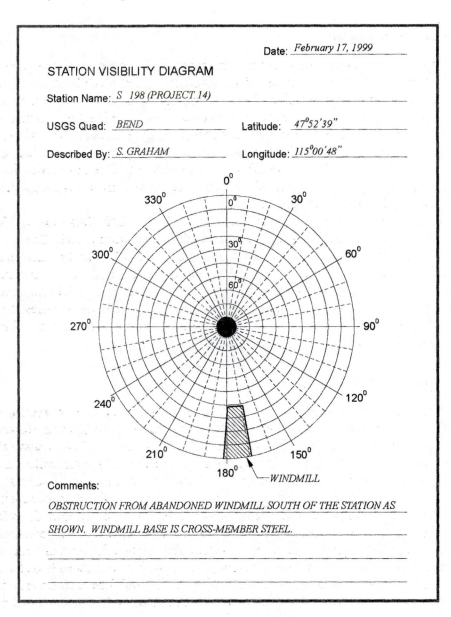

Figure 7.2. Station Visibility Diagram.

of the celestial sphere, from the observer's horizon at 0° on the perimeter, to the observer's zenith at 90° in the center. The hemisphere is cut by the observer's meridian, shown as a line from 0° in the north to 180° on the south. The prime vertical is signified as the line from 90° in the east to 270° in the west. The other numbers and solid lines radiating from the center, every 30° around the perimeter of the figure, are azimuths from north and are augmented by dashed lines every 10°.

Drawing Obstructions

Using a compass and a clinometer, a member of the reconnaissance team can fully describe possible obstructions of the satellite's signals on a visibility diagram. By standing at the station mark and measuring the azimuth and vertical angle of points outlining the obstruction, the observer can plot the object on the visibility diagram. For example, a windmill base is shown in Figure 7.2 as a cross-hatched figure. It has been drawn from the observer's horizon up to 37° in vertical angle from 168°, to about 182° in azimuth at its widest point. This description by approximate angular values is entirely adequate for determining when particular satellites may be blocked at this station.

For example, suppose a 1-hour session from 9:10 to 10:10, illustrated in Table 7.2, was under consideration for the observation on station S 198. The station visibility chart might motivate a careful look at SV PRN 16. Twenty minutes into the anticipated session, at 9:30 SV 16 has just risen above the 15° mask angle. Under normal circumstances, it would be available at station S 198, but it appears from the polar plot that the windmill will block its signals from reaching the receiver. In fact, the signals from SV 16 will apparently not reach station S 198 until sometime after the end of the session at 10:10.

Working around Obstructions

Under the circumstances, some consideration might be given to observing station S 198 during a session when none of the satellites would be blocked. However, the 9:10 to 10:10 session may be adequate after all. Even if SV 16 is completely blocked, the remaining five satellites will be unobstructed and the constellation still will have a relatively low PDOP. Still, the analysis must be carried to other stations that will be occupied during the same session. The success of the measurement of any baseline depends on common observations at both ends of the line. Therefore, if the signals from SV 16 are garbled or blocked from station S 198, any information collected during the same session from that satellite at the other end of a line that includes S 198 will be useless in processing the vector between those two stations.

But the material of the base of the abandoned windmill has been described on the visibility diagram as cross-membered steel, so it is possible that the signal from SV 16 will not be entirely obstructed during the whole session. There may actually be more concern of multipath interference from the structure than that of signal availability. One strategy for handling the situation might be to program the receiver at S 198 to ignore the signal from SV 16 completely if the particular receiver allows it.

The visibility diagram (Figure 7.2) and the azimuth-elevation table (Table 7.2) complement each other. They provide the field supervisor with the data needed to make informed judgments about the observation schedule. Even if the decision is taken to include station S 198 in the 9:10 to 10:10 session as originally planned, the supervisor will be forewarned that the blockage of SV 16 may introduce a bit of weakness at that particular station.

Table 7.2. Satellites' Azimuth and Elevation Table

Time	El	Az	El	Az	El	Az	El	Az	El	Az	El	Az	El	Az	El	Az	PDOP
SV	3		12		13		20		24								
						constellation of 5 SVs											
8:50	54	235	74	274	44	28	16	308	68	169							4.8
9:00	51	229	74	255	40	32	20	310	72	163							5.7
9:10	47	224	72	238	37	35	23	311	77	153							4.9
9:20	43	219	68	226	33	38	27	313	80	134							4.0
SV	3		12		13		16		20		24						
						constellation of 6 SVs											
9:30	39	215	64	218	29	41	16	179	31	314	81	102					2.1
9:40	35	212	59	213	26	45	19	176	36	314	80	73					2.3
9:50	31	209	54	209	23	48	23	173	40	315	76	57					2.4
10:00	27	207	49	206	19	52	27	170	44	314	72	49					2.5
10:10	23	204	44	204	16	55	30	167	48	314	67	45					2.5

Approximate Station Coordinates

The latitude and longitude given on the station visibility diagram should be understood to be approximate. It is sometimes a scaled coordinate, or it may be taken from another source. In either case, its primary role is as input for the receiver at the beginning of its observation. The coordinate need only be close enough to the actual position of the receiver to minimize the time the receiver must take to lock onto the constellation of satellites it expects to find.

Multipath

The multipath condition is by no means unique to GPS. When a transmitted television signal reaches the receiving antenna by two or more paths, the resulting variations in amplitude and phase cause the picture to have ghosts. This kind of scattering of the signals can be caused by reflection from land, water, or man-made structures. In GPS, the problem can be particularly troublesome when signals are received from satellites at low elevation angles; hence the general use of a 15° to 20° mask angle. The use of choke ring antennas to mediate multipath may also be considered.

It is also wise, where it is possible, to avoid using stations that are near structures likely to be reflective or to scatter the signal. For example, chain-link fences that are found hard against a mark can cause multipath by forcing the satellite's signal to pass through the mesh to reach the antenna. The elevation of the antenna over the top of the fence with a survey mast is often the best way to work around this kind of obstruction. Metal structures with large flat surfaces are notorious for causing multipath problems. Offsets well away from such a building are probably the best solution. A long train moving near a project point could be a potential problem, but vehicles passing by on a highway or street usually are not, especially if they go by at high speed. It is important, of course, to avoid parked vehicles. It is best to remind new GPS observers that the survey vehicle should be parked far enough from the point to avert any multipath.

Look for Multipath

Both the GPS field supervisor and the reconnaissance team should be alert to any indications on the station visibility diagram that multipath may be a concern. Before the observations are done, there is nearly always a simple solution. Discovering multipath in the signals after the observations are done is not only frustrating, but often expensive.

Reconnaissance for Kinematic GPS

A careful reconnaissance of the routes between stations is required for successful kinematic GPS. Clear sky for at least five satellites 20° or more above the horizon are a necessity to maintain an adequate lock throughout the survey. Unavoidable obstructions like bridges and tunnels can be overcome by the placement of a control station on both sides of the barrier. If these control stations are coordinated by some type of static observation, they can later be used to reinitialize the kinematic receivers.

Monumentation

The monumentation set for GPS projects varies widely and can range from brass tablets to aerial premarks, capped rebar, or even pin flags. The objective of most station markers is to adequately serve the client's subsequent use. However, the time, trouble, and cost in most high-accuracy GPS work warrants the most permanent, stable monumentation.

Many experts predict that GPS will eventually make monumentation unnecessary. The idea foresees GPS receivers in constant operation at well-known master stations will allow surveyors with receivers to determine highly accurate relative positions with such speed and ease that monumentation will be unnecessary. The idea may prove prophetic, but for now monumentation is an important part of most GPS projects. The suitability of a particular type of monument is an area still left most often to the professional judgment of the surveyors involved.

The FGCC recommends the use of traditional metal disks set in rock outcroppings, bridge abutments, or other large structural elements where possible. A three-dimensional rod mark is approved as an alternative by the federal committee. It is described in detail in Appendix H of *Geometric Geodetic Accuracy Standards and Specifications for Using GPS Relative Positioning Techniques.*

FINAL PLANNING AND OBSERVATION

Logistics

Scheduling

Once all the station data sheets, visibility diagrams, and other field notes have been collected, the schedule can be finalized for the first observations. There will almost certainly be changes from the original plan. Some of the anticipated control stations may be unavailable or obstructed, some project points may be blocked, too difficult to reach, or simply not serve the purpose as well as a control station at an alternate location. When the final control has been chosen, the project points have been monumented, and the reconnaissance has been completed, the information can be brought together with some degree of certainty that it represents the actual conditions in the field.

Now that the access and travel time, the length of vectors, and the actual obstructions are more certainly known, the length and order of the sessions can be solidified. Despite all the care and planning that goes into preparing for a project, unexpected changes in the satellite's orbits or health can upset the best schedule at the last minute. It is always helpful to have a backup plan.

The receiver operators usually have been involved in the reconnaissance and are familiar with the area and many of the stations. Even though an observer may not have visited the particular stations scheduled for him, the copies of the project map, appropriate station data sheets, and visibility diagrams will usually prove adequate to their location.

Observation

When everything goes as planned a GPS observation is uneventful. However, even before the arrival of the receiver operator at the control or project point the session can get off-track. The simultaneity of the data collected at each end of a baseline is critical to the success of any measurement in GPS. When a receiver occupies a master station throughout a project there need be little concern on this subject. But most static applications depend on the sessions of many mobile receivers beginning and ending together.

Arrival

The number of possible delays that may befall an observer on the way to a station are too numerous to mention. With proper planning and reconnaissance, the observer will likely find that there is enough time for the trip from station to station and that sufficient information is on hand to guide him to the position, but this too cannot be guaranteed. When the observer is late to the station, the best course is usually to set up the receiver quickly and collect as much data as possible. The baselines into the late station may or may not be saved, but they will certainly be lost if the receiver operator collects no information at all. It is at times like these that good communication between the members of the GPS team are most useful. For example, some of the other observers in the session may be able to stay on their station a bit longer with the late arrival and make up some of the lost data. Along the same line, it is usually a good policy for those operators who are to remain on a station for two consecutive sessions to collect data as long as possible, while still leaving themselves enough time to reset between the two observation periods.

Setup

Centering an instrument over the station mark is always important. However, the centimeter-level accuracy of static GPS gives the centering of the antenna special significance. It is ironic that such a sophisticated system of surveying can be defeated from such a commonplace procedure. A tribrach with an optical plummet or any other device used for centering should be checked and, if necessary, adjusted before the project begins. With good centering and leveling procedures, an antenna should be within a few millimeters of the station mark. The FGCC's provisional specifications require that the antenna's centering be checked with a plumb bob at each station for surveys of the AA, A, and B orders.

Unfortunately, the centering of the antenna over the station does not ensure that its phase center is properly oriented. The contours of equal phase around the antenna's electronic center are not themselves perfectly spherical. Part of their eccentricity can be attributed to unavoidable inaccuracies in the manufacturing process. To compensate for some of this offset, it is a good practice to rotate all antennas in a session to the same direction. Many manufacturers provide reference marks on their antennas so that each one may be oriented to the same azimuth. That way they are expected to maintain the same relative position between their physical and electronic centers when observations are made.

The antenna's configuration also affects another measurement critical to successful GPS surveying: the height of the instrument. The frequency of mistakes in this important measurement is remarkable. Several methods have been devised to focus special attention on the height of the antenna. Not only should it be measured in both feet and meters, it should also be measured immediately after the instrument is set up and just before tearing it down to detect any settling of the tripod during the observation.

Height of Instrument

The measurement of the height of the antenna in a GPS survey is often not made on a plumb line. A tape is frequently stretched from the top of the station monument to some reference mark on the antenna or the receiver itself. Some GPS teams measure and record the height of the antenna to more than one reference mark on the ground plane. These measurements are usually mathematically corrected to plumb.

The care ascribed to the measurement of antenna heights is due to the same concern applied to centering. GPS has an extraordinary capability to achieve accurate heights, but those heights can be easily contaminated by incorrect H.I.s.

Observation Logs

Most GPS operations require its receiver operators to keep a careful log of each observation. Usually written on a standard form, these field notes provide a written record of the measurements, times, equipment, and other data that explains what actually occurred during the observation itself. It is difficult to overestimate the importance of this information. It is usually incorporated into the final report of the survey, the archives, and any subsequent effort to bluebook the project. However, the most immediate use of the observation log is in evaluation of the day's work by the on-site field supervisor.

An observation log may be organized in a number of ways. The log illustrated in Figure 7.3 is one method that includes some of the information that might be used to document one session at one station. Of course, the name of the observer and the station must be included, and while the date need not be expressed in both the Julian and Gregorian calendars, that information may help in quick cataloging of the data. The approximate latitude, longitude, and height of the station are usually

OBSERVATION LOG

JOB NUMBER [ULY2396]

OBSERVER	STATION	JULIAN DATE	DATE
S. GRAHAM	S 198 (POINT 14)	50	2/17/99

LATITUDE	LONGITUDE	HEIGHT
47°52'39"	115°00'48"	3,241.09 Feet

PLANNED OBSERVATION SESSION	SESSION NAME	ACTUAL OBSERVATION SESSION
START TIME: 9:10 STOP TIME: 10:10	0014 050 2	START TIME: 9:10 STOP TIME: 10:10

ANTENNA TYPE	ANTENNA HEIGHT ABOVE STATION MONUMENT	
	BEFORE OBSERVATION	AFTER OBSERVATION
ON-BOARD	METERS: 1.585	METERS: 1.585
MASK ANGLE: 15°	FEET: 5.20	FEET: 5.20

METEOROLOGICAL DATA				
TIME	RELATIVE HUMIDITY	BAROMETER	THERMOMETER (D)	THERMOMETER (W)
9:30	30%	29.94	37°F	35°F

VISIBILITY DIAGRAM COMPLETED? (Y) N	TOP OF MONUMENT ABOVE THE SURFACE: 3 IN.
STATION DATA SHEET COMPLETED? (Y) N	TOP OF MONUMENT BELOW THE SURFACE:

SV PRN TRACKED		COMMENTS:
3	16	DATA FROM SV 16 APPEARS HEALTHY
12	20	DESPITE WINDMILL OBSERVATION
13	24	

Figure 7.3. Observation Log.

required by the receiver as a reference position for its search for satellites. The date of the planned session will not necessarily coincide with the actual session observed. The observer's arrival at the point may have been late, or the receiver may have been allowed to collect data beyond the scheduled end of the session.

There are various methods used to name observation session in terminology that is sensible to computers. A widely used system is noted here. The first four digits are the project point's number. In this case it is point 14 and is designated 0014. The next three digits are the Julian day of the session; in this case it is day

50, or 050. Finally, the session illustrated is the second of the day, or 2. Therefore, the full session name is 0014 050 2.

Whether onboard or separate, the type of the antenna used and the height of the antenna are critical pieces of information. The relation of the height of the station to the height of the antenna is vital to the station's later utility. The distance that the top of the station's monument is found above or below the surface of the surrounding soil is sometimes neglected. This information can not only be useful in later recovery of the monument, but can also be important in the proper evaluation of photo-control panel points.

Weather

The meteorological data is useful in modeling the atmospheric delay. This information is required at the beginning, middle, and end of each session of projects that are designed to satisfy the FGCC's provisional specifications for the AA and A orders of accuracy. Under those circumstances, measurement of the atmospheric pressure in millibars, the relative humidity, and the temperature in degrees Centigrade are expected to be included in the observation log. However, the general use is less stringent. The conditions of the day are observed, and unusual changes in the weather are noted.

A record of the satellites that are actually available during the observation and any comments about unique circumstances of the session round out the observation log.

Daily Progress Evaluation

The planned observation schedules of a large GPS project usually change daily. The arrangement of upcoming sessions are often altered based on the success or failure of the previous day's plan. Such a regrouping follows evaluation of the day's data.

This evaluation involves examination of the observation logs as well as the data each receiver has collected. Unhealthy data, caused by cycle slips or any other source, are not always apparent to the receiver operator at the time of the observation. Therefore a daily quality control check is a necessary preliminary step before finalizing the next day's observation schedule.

Some field supervisors prefer to actually compute the independent baseline vectors of each day's work to ensure that the measurements are adequate. Neglecting the daily check could leave unsuccessful sessions undiscovered until the survey was thought to be completed. The consequences of such a situation could be expensive.

EXERCISES

1. Of the four listed below what is the most frequent mistake made in GPS control surveying?

    (a) failure to use a 15° mask angle
    (b) failure to have a fully charged battery
    (c) failure to properly measure the H.I. of the antenna
    (d) failure to properly center the antenna over the point

2. Which of the following statements about the zero baseline test is not true?

    (a) The zero baseline text eliminates random noise and receiver biases from the results.
    (b) The zero baseline test requires that receivers share the same antenna using an antenna splitter.
    (c) The zero baseline test eliminates satellite clock biases, ephemeris errors, atmospheric biases, and multipath from the results.
    (d) The zero baseline test is not dependent on special software or a test network.

3. The place of a particular station in a static GPS control survey observation schedule may depend on a number of factors. Which of the following would not be one of them?

    (a) The obstructions around the station.
    (b) The difficulty involved in reaching the station in bad weather.
    (c) The previous day's successful and unsuccessful occupations.
    (d) The line-of-sight with other stations in the survey.

4. The *to-reach* description in a static GPS control survey would most likely appear on which of the following forms?

    (a) The Observation Log
    (b) The Station Data Sheet
    (c) The Station Visibility Diagram
    (d) The Observation Schedule

5. The H.I. of the antenna in a static GPS control survey would most likely appear on which of the following forms?

    (a) The Observation Log
    (b) The Station Data Sheet
    (c) The Station Visibility Diagram
    (d) The Observation Schedule

6. Which of the following forms used in a static GPS control survey would not include information on either the start and stop time of the observation or the approximate latitude and longitude of the station?

 (a) The Observation Log
 (b) The Station Data Sheet
 (c) The Station Visibility Diagram
 (d) The Observation Schedule

7. What is the overriding principle that ought to guide the preparation of a station data sheet?

 (a) to forewarn of obstructed satellites at the site
 (b) to finalize the observation schedule
 (c) to guide succeeding visitors to the station without ambiguity
 (d) to explain what actually occurred during the observation itself

8. Which statement below correctly identifies an advantage of using rubbings in recording the information on an existing monument?

 (a) It ensures that the station was actually visited and that the stamping was faithfully recorded.
 (b) It makes ties to prominent features in the area unnecessary.
 (c) It improves the neatness of the station data sheet.
 (d) It increases the efficiency of the actual occupation of the station.

9. What tools are necessary to prepare a visibility diagram at a station that is somewhat obstructed?

 (a) a theodolite and EDM
 (b) a compass and clinometer
 (c) a GPS receiver
 (d) a level and level rod

10. In planning a kinematic GPS survey, which statement below is the most likely use for control stations placed on either side of a bridge?

 (a) The control stations may be used to reinitialize a receiver that lost lock while passing under the bridge.
 (b) The control stations might be occupied during the survey as a check of the accuracy of the survey.
 (c) The control stations might be used for an antenna swap.
 (d) The control stations might be set to make the kinematic survey unnecessary in the area.

ANSWERS AND EXPLANATIONS

1. Answer is (c)

   Explanation: The mismeasured height of the antenna above the mark is probably the most pervasive and frequent blunder in GPS control surveying.

2. Answer is (a)

   Explanation: The simplicity of the zero baseline test is an advantage. It is not dependent on special software or a test network and it can be used to separate receiver difficulties from antenna errors. Two or more receivers are connected to one antenna with a *signal,* or *antenna, splitter.* An observation is done with the divided signal from the single antenna reaching both receivers simultaneously. Since the receivers are sharing the same antenna, satellite clock biases, ephemeris errors, atmospheric biases, and multipath are all canceled. The only remaining errors are attributable to random noise and receiver biases.

3. Answer is (d)

   Explanation: In creating an observation schedule, consideration might be given to observing a particular station during a session when none of the satellites would be blocked by obstructions. And while GPS is not restricted by inclement weather, particular access routes may not be so immune. Despite best efforts, a planned observation may have been unsuccessful at a required station on a previous day, and it may need to be revisited. However, the line-of-sight between a particular station and another in the survey is not likely to affect the GPS observation schedule, though such a consideration may be critical in a conventional survey.

4. Answer is (b)

   Explanation: The *to-reach* description in a static GPS control survey would most likely be prepared during reconnaissance and would appear on the station data sheet.

5. Answer is (a)

   Explanation: The H.I. of the antenna in a static GPS control survey would most likely be recorded before and after the actual observation on the station and would appear on the observation log.

6. Answer is (b)

   Explanation: The station data sheet would not be likely to include either information. The other forms listed would probably include one or both categories.

7. Answer is (c)

   Explanation: The station data sheet is often an important bridge between on-site reconnaissance and the actual occupation of a monument. Neatness and clarity, always paramount virtues of good field notes, are of particular interest when the station data sheet is to be included later in the final report to the client. The overriding principle in drafting a station data sheet is to guide succeeding visitors to the station without ambiguity. A GPS surveyor on the way to observe the position for the first time may be the initial user of a station data sheet. A poorly written document could void an entire session if the observer is unable to locate the monument. A client, later struggling to find a particular monument with an inadequate data sheet, may ultimately question the value of more than the field notes.

8. Answer is (a)

   Explanation: Rubbings are performed with paper held on top of the monument's disk; a pencil is run over it in a zigzag pattern producing a positive image of the stamping. This method is a bit more awkward than simply copying the information from the disk onto the data sheet, but it does have the advantage of ensuring the station was actually visited and that the stamping was faithfully recorded. Such rubbings or close-up photographs are required by the provisional FGCC *Geometric Geodetic Accuracy Standards and Specifications for Using GPS Relative Positioning Techniques* for all orders of GPS surveys.

9. Answer is (b)

   Explanation: Using a compass and a clinometer, a member of the reconnaissance team can fully describe possible obstructions of the satellite's signals on a visibility diagram. By standing at the station mark and measuring the azimuth and vertical angle of points outlining the obstruction, the observer can plot the object on the visibility diagram.

10. Answer is (a)

    Explanation: Unavoidable obstructions like bridges and tunnels can be overcome by the placement of control station on both sides of the barrier. If these control stations are coordinated by some type of static observation, they can later be used to reinitialize the kinematic receivers should they lose lock passing under the bridge.

8

Postprocessing

PROCESSING

In many ways, processing is the heart of a GPS operation. Some processing should be performed on a daily basis during a GPS project. Blunders from operators, noisy data, and unhealthy satellites can corrupt entire sessions. And left undetected, such dissolution can jeopardize an entire survey. But with some daily processing, these weaknesses in the data can be discovered when they can still be eliminated with a timely amendment of the observation schedule.

But even after blunders and noisy data have been removed from the observation sets, GPS measurements are still composed of fundamentally biased ranges. Therefore, GPS data-processing procedures are really a series of interconnected computerized operations designed to remove these more difficult biases and extract the true ranges.

The biases originate from a number of sources: imperfect clocks, atmospheric delays, cycle ambiguities in carrier phase observations, and orbital errors. If a bias has a stable, well-understood structure, it can be estimated. In other cases, dual-frequency observations can be used to measure the bias directly, as in the ionospheric delay, or a model may be used to predict an effect, as in tropospheric delay. But one of the most effective strategies in eradicating biases is called *differencing*.

Correlation of Biases

When two or more receivers observe the same satellite constellation simultaneously, a set of correlated vectors are created between the co-observing stations. Most GPS practitioners use more than two receivers. Therefore, most GPS networks consist of many sets of correlated vectors for every separate session. The longest baselines between stations on the earth are usually relatively short when compared with the more than 20,000-km distances from the receivers to the GPS satellites. Therefore, even when several receivers are set up on widely spaced stations, as long as they collect their data simultaneously from the same constellation of satellites, they will record very similar errors. In other words, their vectors will be correlated. It is the simultaneity of observation and the resulting correlation of the carrier phase observables that make the extraordinary GPS accuracies possible. Biases that are correlated linearly can be virtually eliminated by differencing the data sets of a session.

Quantity of Data

Organization Is Essential

One of the difficulties of GPS processing is the huge amount of data that must be managed. For example, when even one single-frequency receiver with a 1-second sampling rate tracks one GPS satellite for an hour, it collects about 0.15 Mb of data. However, a more realistic scenario involves four receivers observing six satellites for 3,600 epochs. There can be $4 \times 6 \times 3,600$ or 86,400 carrier phase observations in such a session. In other words, a real-life GPS survey with many sessions and many baselines creates a quantity of data in the gigabyte range. Some sort of structured approach must be implemented to process such a huge amount of information in a reasonable amount of time.

File Naming Conventions

One aspect of that structure is the naming conventions used to head GPS receivers' observation files. Many manufacturers recommend a file naming format that can be symbolized by *pppp-ddd-s.yyf*. The first letters, (*pppp*), of the file name indicate the point number of the station occupied. The day of the year, or Julian date, can be accommodated in the next three places (*ddd*), and the final place left of the period is the session number (*s*). The year (*yy*), and the file type (*f*) are sometimes added to the right of the period.

Downloading

The first step in GPS data processing is downloading the collected data from the internal memory of the receiver itself into a PC or laptop computer. When the observations sessions have been completed for the day, each receiver, in turn, is cabled to the computer and its data transferred. Nearly all GPS systems used in surveying are PC-compatible and can accommodate postprocessing in the field. But none can protect the user from a failure to back up this raw observational data onto some other form of semipermanent storage.

Making Room

Receiver memory capacity is usually somewhat limited, and older data must be cleared to make room for new sessions. Still, it is a good policy to create the necessary space with the minimum deletion and restrict it to only the oldest files in the receiver's memory. In this way the recent data can be retained as long as possible, and the data can provide an auxiliary backup system. But when a receiver's memory is finally wiped of a particular session, if redundant raw data are not available reobservation may be the only remedy.

Most GPS receivers record data internally. The Navigation message, meteorological data, the observables, and all other raw data are usually in a manufacturer-specific, binary form. These raw data are usually saved in several distinct files. For example, the PC operator will likely find that the phase measurements downloaded from the receiver will reside in one file and the satellite's ephemeris data in another. Likewise, the measured pseudorange information may be found in its own dedicated file, the ionospheric information in another, and so on. The particular division of the raw data files is designed to accommodate the suite of processing software and the data management system that the manufacturer has provided its customers, so each will be somewhat unique.

Control

All postprocessing software suites require control. Control is usually entered in the degrees, minutes, and seconds of latitude and longitude and orthometric height, but most will also accept other formats.

The First Position

Baseline processing is usually begun with a point position solution at each end from pseudoranges. These differential code estimations of the approximate position of each receiver antenna can be thought of as establishing a search area, a three-dimensional volume of uncertainty at each receiver containing its correct position. The size of this search area is defined by the accuracy of the code solution, which also affects the computational time required to find the correct position among all the other potential solutions.

Triple Difference

The next step, usually the triple-difference, utilizes the carrier phase observable. Triple differences have several features to recommend them for this stage in the processing. They can achieve rather high accuracy even before cycle slips have been eliminated from the data sets, and they are insensitive to integer ambiguities in general.

Components of a Triple Difference

A triple difference is created by differencing two double differences at each end of the baseline. Each of the double differences involves two satellites and two receivers. A triple difference considers two double differences over two consecutive epochs. In other words, triple differences are formed by sequentially differencing double differences in time. For example, two triple differences can be created using double differences at epochs 1, 2, and 3. One is double difference 2 minus double difference 1. A second can be formed by double difference 3 minus double difference 2.

Since two receivers are recording the data from the same two satellites during two consecutive epochs across a baseline, a triple difference can temporarily eliminate any concern about the integer cycle ambiguity, because the cycle ambiguity is the same over the two observed epochs. However, the triple difference cannot have as much information content as a double difference. Therefore, while receiver coordinates estimated from triple differences are usually more accurate than pseudorange solutions, they are less accurate than those obtained from double differences, especially fixed-ambiguity solutions. Nevertheless, the estimates that come out of triple-difference solutions refine receiver coordinates and provide a starting point for the subsequent double-difference solutions. They are also very useful in spotting and correcting cycle slips. They also provide a first estimate of the receiver's positions.

Double Difference

The next baseline processing steps usually involve two types of double differences, called the *float* and the *fixed* solutions.

The Integer Ambiguity

Double differences have both positive and negative features. On the positive side, they make the highest GPS accuracy possible, and they remove the satellite and receiver clock errors from the observations. On the other hand, the integer cycle ambiguity, sometimes known simply as *the ambiguity,* cannot be ignored in the double difference. In fact, the fixed double-difference solution, usually the most accurate technique of all, requires the resolution of this ambiguity.

The integer cycle ambiguity, usually symbolized by N, represents the number of full phase cycles between the receiver and the satellite at the first instant of the receiver's lock-on. N does not change from the moment the lock is achieved, unless there is a cycle slip. Unfortunately, N is also an unknown quantity at the beginning of any carrier phase observation.

The Float Solution

Once again estimation plays a significant role in finding the appropriate integer value that will correctly resolve the ambiguity for component pairs in double differencing. In this first try, there is no effort to translate the biases into integers. It is sometimes said that the integers are allowed to float; hence the initial process is called the *float solution.* Especially when phase measurements for only one frequency, L1 or L2, are available, a sort of calculated guess at the ambiguity is the most direct route to the correct solution. Not just N, but a number of unknowns, such as clock parameters and point coordinates, are estimated in this geometric approach. However, all of these estimated biases are affected by unmodeled errors, and that causes the integer nature of N to be obscured. In other words, the initial estimation of N in a float solution is likely to appear as a real number rather than an integer.

However, where the data are sufficient, these floating real-number estimates are very close to integers—so close that they can next be rounded to their true integer values in a second adjustment of the data. Therefore, a second double difference solution follows. The estimation of N that is closest to an integer and has the minimum standard error is usually taken to be the most reliable and is rounded to the nearest integer. Now, with one less unknown, the process is repeated and another ambiguity can be fixed, and so on.

The Fixed Solution

This approach leads to the *fixed solution* in which N can be held to integer values. It is usually quite successful in double differences over short baselines. The resulting fixed solutions most often provide much more accurate results than were available from the initial floating estimates.

Cycle Slip Detection and Repair

However, this process can be corrupted by the presence of cycle slips. A cycle slip is a discontinuity in a receiver's continuous phase lock on a satellite's signal. The coded pseudorange measurement is immune from this difficulty, but the carrier beat phase is not. In other words, even when a fixed double-difference solution can provide the correct integer ambiguity resolution, the moment the data set is interrupted by a cycle slip, that solution is lost.

There are actually two components to the carrier phase observable that ought not to change from the moment of a receiver's lock onto a particular satellite. First is the fractional initial phase at the first moment of the lock. The receiver is highly unlikely to acquire the satellite's signal precisely at the beginning of a wavelength. It will grab on at some fractional part of a phase, and this fractional phase will remain unchanged for the duration of the observation. The other unchanged aspect of a normal carrier phase observable is an integer number of cycles. The integer cycle ambiguity is symbolized by N. It represents the number of full phase cycles between the receiver and the satellite at the first instant of the receiver's lock-on. The integer ambiguity ought to remain constant throughout an observation as well. But when there is a cycle slip, lock is lost, and by the time the receiver reacquires the signal, the normally constant integer ambiguity has changed.

Cycle Slip Causes

A power loss, a very low signal-to-noise ratio, a failure of the receiver software, a malfunctioning satellite oscillator, or any event that breaks the receiver's continuous reception of the satellite's signal causes a cycle slip. Most common, however, is an obstruction that is so solid it prevents the satellite signal from being tracked by the receiver. Under such circumstances, when the satellite reappears, the tracking resumes. The fractional phase may be the same as if tracking had been maintained, but the integer number of cycles is not.

Repairing Cycle Slips

Cycle slips are repaired in postprocessing. Both their location and their size must be determined; then the data set can be repaired with the application of a fixed quantity to all the subsequent phase observations. One approach is to hold the initial positions of the stations occupied by the receivers as fixed and edit the data manually. This has proved to work, but would try the patience of Job. Another approach is to model the data on a satellite-dependent basis with continuous polynomials to find the breaks and then manually edit the data set a few cycles at a time. In fact, several methods are available to find the lost integer phase value, but they all involve testing quantities.

One of the most convenient of these methods is based on the triple difference. It can provide an automated cycle slip detection system that is not confused by clock drift and once least-squares convergence has been achieved, it can provide initial station positions even using the unrepaired phase combinations. They may still contain cycle slips but can nevertheless be used to process approximate baseline vectors. Then the residuals of these solutions are tested, sometimes through several iterations. Proceeding from its own station solutions, the triple difference can predict how many cycles will occur over a particular time interval. Therefore, by evaluating triple-difference residuals over that particular interval, it is not only possible to determine which satellites have integer jumps, but also the number of cycles that have actually been lost. In a sound triple-difference solution without cycle slips, the residuals are usually limited to fractions of a cycle. Only those containing cycle slips have residuals close to one cycle or larger. Once cycle slips are discovered, their correction can be systematic.

For example, suppose the residuals of one component double difference of a triple-difference solution revealed that the residual of satellite PRN 16 minus the residual of satellite PRN 17 was 8.96 cycles. Further suppose that the residuals from the second component double difference showed that the residual of satellite PRN 17 minus the residual of satellite PRN 20 was 14.04 cycles. Then one might remove 9 cycles from PRN 16 and 14 cycles from PRN 20 for all the subsequent epochs of the observation. However, the process might result in a common integer error for PRNS 16, 17, and 20. Still, small jumps of a couple of cycles can be detected and fixed in the double difference solutions.

In other words, before attempting double-difference solutions, the observations should be corrected for cycle slips identified from the triple difference solution. And even though small jumps undiscovered in the triple difference solution might remain in the data sets, the double difference residuals will reveal them at the epoch where they occurred.

However, some conditions may prevent the resolution of cycle slips down to the one-cycle level. Inaccurate satellite ephemerides, noisy data, errors in the receiver's initial positions, or severe ionospheric effects all can limit the effectiveness of cycle-slip fixing. In difficult cases, a detailed inspection of the residuals might be the best way to locate the problem.

Fixing the Integer Ambiguity and Obtaining Vector Solutions

When cycle slips have been finally eliminated, the ambiguities can be fixed to integer values. First, the standard deviations of the adjusted integers are inspected. If found to be significantly less than one cycle, they can be safely constrained to the nearest integer value. This procedure is only pertinent to double differences, since phase ambiguities are moot in triple differencing. The number of parameters involved can be derived by multiplying one less the number of receivers by one less the number of satellites involved in the observation. And for dual-frequency observations the phase ambiguities for L1 and L2 are best fixed separately. But such a program of constraints is not typical in long baselines where the effects of the ionosphere and inaccuracies of the satellite ephemerides make the situation less determinable.

In small baselines, however, where these biases tend to be virtually identical at both ends, integer fixing is almost universal in GPS processing. Once the integer ambiguities of one baseline are fixed, the way is paved for constraint of additional ambiguities in subsequent iterations. Then, step by step, more and more integers are set, until all that can be fixed have been fixed. Baselines of several thousand kilometers can be constrained in this manner.

With the integer ambiguities fixed, the GPS observations produce a series of vectors, the raw material for the final adjustment of the survey. The observed baselines represent very accurately determined relative locations between the stations they connect. However, the absolute position of the whole network is usually much less accurately known, although more accurate absolute positioning may be on the horizon for GPS. For now, there remains a considerable difference in the accuracy of relative and absolute positioning. A GPS survey is usually related to the rest of the world by translation of its Cartesian coordinates into ellipsoidal latitude, longitude, and height. Most users are less comfortable with the original coordinate results, given in the WGS84 coordinate system of the satellites themselves, than latitude, longitude, and height.

Adjustment

Least-Squares

There are numerous adjustment techniques, but least-squares adjustment is the most precise and most commonly used in GPS. The foundation of the idea of least-squares adjustment is the idea that the sum of the squares of all the residuals applied to the GPS vectors in their final adjustment should be held to the absolute minimum. But minimizing this sum requires first defining those residuals approximately. Therefore, in GPS it is based on equations where the observations are expressed as a function of unknown parameters, but parameters that are nonetheless given approximate initial values. This is the process that has been described above. Then by adding the squares of the terms thus formed and differentiating their sum, the derivatives can be set equal to zero. For complex work like GPS adjustments, the least-squares method has the advantage that it allows for the smallest possible changes to original estimated values.

The solution strategies of GPS adjustments themselves are best left to particular suites of software. Suffice it to say that the single baseline approach, that is, a baseline-by-baseline adjustment, has the disadvantage of ignoring the actual correlation of the observations of simultaneously occupied baselines. An alternative approach involves a network adjustment approach where the correlation between the baselines themselves can be more easily taken into account. And while the computations are simpler for the baseline-by-baseline approach, cycle slips are more conveniently repaired in network adjustment.

For the most meaningful network adjustment, the endpoints of every possible baseline should be connected to at least two other stations. Thereby, the quality of the work itself can be more realistically evaluated. For example, the most common observational mistake, the mismeasured antenna height, is very difficult to detect when adjusting baselines sequentially, one at a time, but a network solution spots such blunders more quickly.

For example, most GPS postprocessing adjustment begins with a minimally constrained least-squares adjustment. That means that all the observations in a network are adjusted together with only the constraints necessary to achieve a meaningful solution. The adjustment of a GPS network with the coordinates of only one station fixed. The purpose of the minimally constrained approach is to detect large mistakes, like misidentifying one of the stations. The residuals from a minimally constrained work should come pretty close to the precision of the observations themselves. If the residuals are particularly large, there are probably mistakes; if they are really small, the network itself may not be as strong as it should be.

This minimally constrained solution is usually followed by an overconstrained solution. In other words, a least-squares adjustment where the coordinate values of selected control station are held fixed.

It is important to note that the downside of least-squares adjustment is its tendency to spread the effects of even one mistake throughout the work. In other words, it can cause large residuals to show up for several measurements that are actually correct. And when that happens it can be hard to know exactly what is wrong. The adjustment may fail the *chi-square* test. That tells you there is a problem; unfortunately it cannot tell you where the problem is. The chi square test is based in probability, and it can fail because there are still unmodeled biases in the measurements. Multipath, ionosphere and troposphere biases, etc., may cause it to fail.

Most programs look at the residuals in light of a limit at a specific probability, and when a particular measurement goes over the limit it gets highlighted. Trouble is, you cannot always be sure that the one that got tagged is the one that is the problem. Fortunately, least-squares does offer a high degree of comfort once all the hurdles have been cleared. If the residuals are within reason and the chi-square test is passed, it is very likely that the observations have been adjusted properly.

EXERCISES

1. Which of the following statements concerning the correlation of biases in GPS is not correct?

   (a) Even relatively long GPS baselines are short when compared with the more than 20,000-km distances from the receivers to the GPS satellites.
   (b) Every session included in a network is composed of many sets of vectors with correlated errors at each end.
   (c) The simultaneity of observation and the resulting correlation of errors helps make the extraordinary accuracy of GPS possible.
   (d) The effect of correlated errors in GPS baselines cannot be reduced by differencing.

2. Should the memory of GPS receivers be cleared routinely after their observations are downloaded to the memory of the processing computer? Why, or why not?

   (a) Yes, the receiver's memory is limited and should be made available for subsequent work.
   (b) No, it is always wise to have a copy of the receiver's observation in the receiver as long as possible as a backup.
   (c) Yes, once the file has been downloaded there is no reason to keep it in the receiver as well.
   (d) No, it is not possible to download critical information from the receiver's files to the processing computer.

3. Which statement concerning triple differences is most correct?

   (a) A triple difference is the difference between the carrier-phase observations of two receivers of the same satellite at the same epoch.
   (b) A triple difference is the difference of two single differences of the same epoch that refer to two different satellites.
   (c) A triple differences is the difference of two double differences of two different epochs.
   (d) A triple difference does depend on the integer cycle ambiguity, because this unknown is not constant in time.

4. What is the most useful aspect of a triple difference in postprocessing GPS observations?

   (a) Triple differences can be an initial step to eliminate cycle slips.
   (b) Triple differences provide the best solution for the receiver positions.
   (c) Triple differences have more information than double differences.
   (d) Triple differences eliminate the need to model ionospheric biases.

5. What is the most useful aspect of a double difference in postprocessing GPS observations?

   (a) Double differences don't need resolution of the integer ambiguity.
   (b) Double differences eliminate clock errors.
   (c) Double differences eliminate multipath.
   (d) Double differences eliminate all atmospheric biases.

6. What would cause the integer cycle ambiguity, N, to change after a receiver has achieved lock-on?

   (a) an incorrect H.I.
   (b) inclement weather
   (c) inaccurate centering over the station
   (d) a cycle slip

7. What is the difference between a float and a fixed solution?

   (a) The integer cycle ambiguity is resolved in a fixed solution, but not in a float solution.
   (b) The float solution is processed using a single difference, unlike a fixed solution.
   (c) The float solution is more accurate than a fixed solution.
   (d) The float solution may have cycle slips, but not a fixed solution.

8. What is the idea underlying the use of least-squares adjustment of GPS networks?

   (a) The sum of the squares of all the residuals applied to the GPS vectors in their final adjustment should be held to zero.
   (b) The multiplication of the GPS vectors by all the residuals in their final adjustment should be held to one.
   (c) The sum of the squares of all the residuals applied to the GPS vectors in their final adjustment should be held to absolute minimum.
   (d) The least of the squares of all the residuals applied to the GPS vectors in their final adjustment should be held to zero.

9. What is usually the purpose of the initial minimally constrained least-squares adjustment in GPS work?

   (a) to repair cycle slips
   (b) to fix the integer cycle ambiguity
   (c) to establish the correct coordinates of all the project points
   (d) to find any large mistakes in the work

10. What may cause an adjustment to fail the chi-square test?

   (a) unmodeled biases in the measurements
   (b) smaller than expected residuals
   (c) a measurement with too many epochs
   (d) the lack of cycle slips

ANSWERS AND EXPLANATIONS

1. Answer is (d)

   Explanation: When two or more receivers observe the same satellite constellation simultaneously, a set of correlated vectors are created between the co-observing stations. Most GPS practitioners use more than two receivers. Therefore, most GPS networks consist of many sets of correlated vectors for every separate session. The longest baselines between stations on the earth are usually relatively short when compared with the 20,000-km distances from the receivers to the GPS satellites. Therefore, even when several receivers are set up on widely spaced stations, as long as they collect their data simultaneously from the same constellation of satellites, they will record very similar errors. In other words, their vectors will be correlated. It is the simultaneity of observation and the resulting correlation of the carrier phase observables that make the extraordinary GPS accuracies possible. Biases that are correlated linearly can be virtually eliminated by differencing the data sets of a session.

2. Answer is (b)

   Explanation: Receiver memory capacity is usually somewhat limited, and older data must be cleared to make room for new sessions. Still, it is a good policy to create the necessary space with the minimum deletion and restrict it to only the oldest files in the receiver's memory. In this way the recent data can be retained as long as possible; the data can provide an auxiliary backup system. But when a receiver's memory is finally wiped of a particular session, if redundant raw data are not available, reobservation may be the only remedy.

3. Answer is (c)

Explanation: A triple difference is created by differencing two double differences at each end of the baseline. Each of the double differences involves two satellites and two receivers. A triple difference considers two double differences over two consecutive epochs. In other words, triple differences are formed by sequentially differencing double differences in time.

Since two receivers are recording the data from the same two satellites during two consecutive epochs across a baseline, a triple difference can temporarily eliminate any concern about the integer cycle ambiguity, because the cycle ambiguity is the same over the two observed epochs.

4. Answer is (a)

Explanation: The triple difference cannot have as much information content as a double difference. Therefore, while receiver coordinates estimated from triple differences are usually more accurate than pseudorange solutions, they are less accurate than those obtained from double differences, especially fixed-ambiguity solutions. Nevertheless, the estimates that come out of triple-difference solutions refine receiver coordinates and provide a starting point for the subsequent double-difference solutions. They are also very useful in spotting and correcting cycle slips. They also provide a first estimate of the receiver's positions.

5. Answer is (b)

Explanation: Double differences have both positive and negative features. On the positive side, they make the highest GPS accuracy possible, and they remove the satellite and receiver clock errors from the observations. On the other hand, the integer cycle ambiguity, sometimes known simply as *the ambiguity,* cannot be ignored in the double difference. In fact, the fixed double-difference solution, usually the most accurate technique of all, requires the resolution of this ambiguity.

6. Answer is (d)

Explanation: The integer cycle ambiguity, usually symbolized by N, represents the number of full phase cycles between the receiver and the satellite at the first instant of the receiver's lock-on. N does not change from the moment the lock is achieved, unless there is a cycle slip.

7. Answer is (a)

Explanation: In other words, the initial estimation of N in a float solution is likely to appear as a real number rather than an integer. However, where the data are sufficient, these floating real-number estimates are very close to integers—so close that they can next be rounded to their true integer values in a second adjustment of the data. Therefore, a second double difference solution follows. The estimation of N that is closest to an integer and has the minimum standard error is usually taken to be the most reliable and is rounded to the nearest integer. Now, with one less unknown, the process is repeated and another ambiguity can be fixed, and so on. This approach leads to the fixed solution in which N can be held to integer values. It is usually quite successful in double differences over short baselines. The resulting fixed solutions most often provide much more accurate results than were available from the initial floating estimates.

8. Answer is (c)

Explanation: The foundation of the idea of least-squares adjustment is the idea that the sum of the squares of all the residuals applied to the GPS vectors in their final adjustment should be held to the absolute minimum.

9. Answer is (d)

Explanation: Most GPS postprocessing adjustment begins with a minimally constrained least-squares adjustment. That means that all the observations in a network are adjusted together with only the constraints necessary to achieve a meaningful solution. For example, the adjustment of a GPS network with the coordinates of only one station fixed. The purpose of the minimally constrained approach is to detect large mistakes, like misidentifying one of the stations. The residuals from a minimally constrained work should come pretty close to the precision of the observations themselves. If the residuals are particularly large, there are probably mistakes; if they are really small, the network itself may not be as strong as it should be.

10. Answer is (a)

Explanation: The adjustment may fail the *chi-square* test. That tells you there is a problem; unfortunately it cannot tell you where the problem is. The chi square test is based in probability, and it can fail because there are still unmodeled biases in the measurements. Multipath, ionosphere, and troposphere biases, etc., may cause it to fail.

Most programs look at the residuals in light of a limit at a specific probability, and when a particular measurement goes over the limit it gets highlighted. Trouble is, you cannot always be sure that the one that got tagged is the one that is the problem. Fortunately, least-squares does offer a high degree of comfort once all the hurdles have been cleared. If the residuals are within reason and the chi-square test is passed, it is very likely that the observations have been adjusted properly.

9

RTK and DGPS

NEW DEVELOPMENTS

It is abundantly clear that survey-grade GPS technology is vastly improved. Receivers are lighter, less expensive, and it takes less time to get good results. For example, in the early 1980s, not only did receivers have to simultaneously track the carrier phase for hours to get a good static position, but processing software was cumbersome and the satellites were sparse. It was difficult. But, even then, GPS provided such remarkable results that overcoming the obstacles was completely justified.

Today, a surveyor seldom has to spend the wee hours at a single station with an extremely expensive GPS receiver. Things have changed. In fact, GPS is now so widely used that the surveyor using a GPS receiver will find there is a good chance passersby will have a pretty good idea what he or she is doing. Not so long ago, that would not have been very likely. But even with all the improvements in GPS applications many surveyors are reluctant to take the plunge. Total stations and more conventional methods are reliable, consistent, and proven producers. All of these developments may explain to some degree the growth in the use of GPS methods that are stable, fast, accurate, and well suited to everyday surveying projects, *real-time kinematic GPS,* also known as *RTK* and *Differential GPS,* also known as *DGPS.*

THE GENERAL IDEA

Errors in satellite clocks, imperfect orbits, the trip through the layers of the atmosphere, and many other sources contribute inaccuracies to GPS signals by the time they reach a receiver. These errors are variable, so the best way to correct them is to monitor them as they happen.

A good way to do this is to set up a second GPS receiver on a station whose position is known exactly, a *base station.* This base station can calculate its position from satellite data, compare that position with its actual known position, find the difference and presto, error corrections. It works well, but the errors are constantly changing so that base station has to monitor them all the time, at least all the time the *rover receiver* or receivers are working. The rover receiver, or more than one roving receiver, moves from place to place collecting the points whose positions you want to know relative to the base station, which is the real objective after all. And it is worth remembering that the real strength of GPS is its ability to provide relative accuracy. In that case, all you have to do is get those base station corrections and that rover's data together somehow. It can be done with a radio link, and thereby apply corrections as the rover works, in real-time, or the corrections can be applied some time later, in postprocessing.

In other words, the basic idea is that by having a base station at a known location you can improve the accuracy of a roving GPS receiver, or a whole bunch of roving receivers for that matter. Both RTK and DGPS have been built on this foundation.

DGPS

The term DGPS is most often used to refer to differential GPS that is based on pseudoranges. The most fundamental configuration for DGPS uses one receiver as the base; that is the reference or monitoring station, and one rover (Figure 9.1). However, there can be several rovers.

Identical Constellation

As mentioned above, this technique requires that all receivers collect pseudoranges from the same constellation of satellites. It is vital that the errors corrected by the base station are common to the rovers. The rover must share its selection of satellites with the base station, otherwise the base station would be obliged to create differential corrections for all the combinations of all the available satellites. That could get unmanageable in a hurry.

Latency

Also, it takes some time for the base station to calculate these errors and it takes some time for it to transmit them. The base station's data is put into packets in the correct format. The data makes its way from the base station to the rover over the data link. It must then be decoded and go through the rover's software. The time it takes for all of this to happen is called the *latency* of the communication between the base station and the rover. It can be as little as a quarter of a second or as long as a couple of seconds. And since the base station's corrections are only accurate for the moment they were created, the base station must send a range rate correction along with them. Using this rate correction, the rover can *back date* the correction to match the moment it made that same observation.

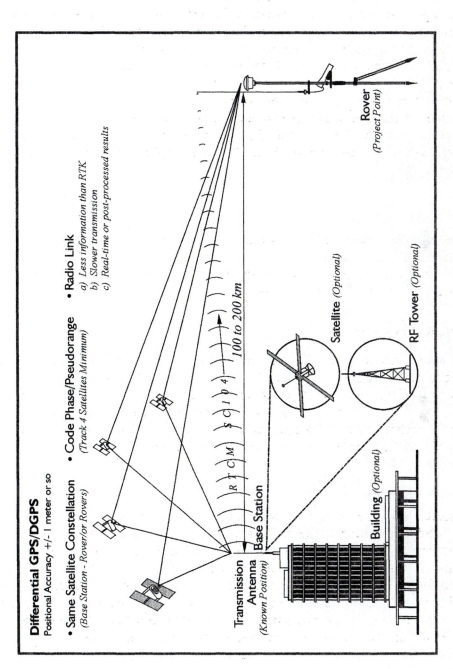

Differential GPS/DGPS
Positional Accuracy +/- 1 meter or so

- **Same Satellite Constellation**
  (Base Station - Rover/or Rovers)

- **Code Phase/Pseudorange**
  (Track 4 Satellites Minimum)

- **Radio Link**
  a) *Less information than RTK*
  b) *Slower transmission*
  c) *Real-time or post-processed results*

100 to 200 km

Transmission
Antenna
(Known Position)

Base Station

Building *(Optional)*

Satellite *(Optional)*

RF Tower *(Optional)*

Rover
(Project Point)

Figure 9.1. DGPS.

Real-Time or Postprocessed

The biases calculated by the base station can reduce the range of common errors, or *correlated errors,* with its differential corrections. If results are required in real-time, the base station transmits these differential corrections via a radio link to the rovers. In that case, the rovers can calculate earth-centered-earth-fixed coordinates using these corrections, or alternatively the rover and the base station store the observations and the results are postprocessed.

Error Correlation and Accuracy

As mentioned above, DGPS improves on autonomous GPS positioning by reducing the effect of correlated errors from two or more receivers only as long as they are all observing the same satellites. Horizontal position accuracies of ±1 m (2 drms) are achievable. Of course, even better accuracy is possible if the carrier phase observable is used in the same configuration; more about that later.

DGPS accuracy depends on the fact that many of the pseudorange biases are virtually the same for all receivers in the vicinity of the base station. Of course, the farther a rover stands from the base station, the less its errors are correlated with the base station, and the less accurate its position will be. In autonomous point positioning all the biases discussed earlier affect the GPS position. For example, in autonomous point positioning the satellite clock bias normally contributes about 3 m to the final GPS positional error, ephemeris bias contributes about 5 m and solar radiation, etc., contributes about another 1.5 m. But with DGPS, when errors are well correlated all these biases are virtually zero, as are the atmospheric errors, unlike autonomous point positioning where the ionospheric delay contributes about 5 m and the troposphere about 1.5 m.

In short, along with other miscellaneous biases an autonomous point position in GPS with SA turned off will usually have about ±20 m to ±40 m of systematic bias in a horizontal position, whereas a DGPS position may be subject to a tenth of that, or even less. It is important to remember, however, that there are other errors that are certainly not eliminated in DGPS such as receiver noise and multipath.

Real-Time DGPS

DGPS corrections are available in a format known as *RTCM.* These corrections are available from a couple of sources, some are free and some by subscription. With these corrections it has become possible to perform DGPS surveys without your own base station. The RTCM message can provide the necessary corrections in real-time.

RTCM-104 MESSAGE FORMAT

In DGPS there is an agreed-upon format for the communication between the base station and rovers. It was first designed for marine navigation. In 1985, *the*

Radio Technical Commission for Maritime Services, RTCM, Study Committee, SC-104, created a standard format for DGPS corrections that is still more used than any proprietary formats that have come along since. Both the United States and the Canadian Coast Guards use this format in their coastal beacons.

This system of beacons are base stations that broadcast GPS corrections along major rivers, major lakes, the east coast and the west coast. Access to the broadcasts are free. But there are also subscription services that broadcast RTCM corrections by satellite and still others that use tower-mounted transmitters (Figure 9.1). In all cases, the base stations are at known locations and their corrections are broadcast to all rovers that are equipped to receive their particular radio message carrying real-time corrections in the RTCM format.

GIS APPLICATIONS FOR DGPS

Aerial navigation, marine navigation, agriculture, vehicle tracking, and construction are all now using DGPS. DGPS is also useful in land and hydrographic surveying, but perhaps the fastest growing application for DGPS is in data collection, data updating, and even in-field mapping for Geographic Information Systems, GIS.

GIS data has long been captured from paper records such as digitizing and scanning paper maps. Photogrammetry, remote sensing, and conventional surveying have also been data sources for GIS. More recently, data collected in the field with DGPS have become significant in GIS. GIS data collection with DGPS requires the integration of the position of features of interest and relevant attribute information about those features. Whether a handheld datalogger, an electronic notebook or a pen computer are used, the attributes to be collected are defined by the data dictionary designed for the particular GIS.

In GIS it is frequently important to return to a particular site or feature to perform inspections or maintenance. DGPS with real-time correction makes it convenient to load the position or positions of features into a datalogger, and navigate back to the vicinity. But to make such applications feasible, a GIS must be kept current. It must be maintained. A receiver configuration including real-time DGPS, sufficient data storage and graphic display allows easy verification and updating of existing information.

DGPS allows the immediate attribution and validation in the field with accurate and efficient recording of position. In the past many GIS mapping efforts have often relied on ties to street centerlines, curb-lines, railroads and etc. Such dependencies can be destroyed by demolition or new construction. But meter level positional accuracy even in obstructed environments such as urban areas, amid high-rise building is possible with DGPS. In other words, with DGPS the control points are the satellites themselves, therefore it can provide reliable positioning even if the landscape has changed. And its data can be integrated with other technologies, such as laser range-finders, etc. in environments where DGPS is not ideally suited to the situation.

Finally, loading GPS data into a GIS platform does not require manual intervention. GPS data processing can be automated, the results are digital and can pass into a GIS format without redundant effort, reducing the chance for errors.

RTK

Kinematic surveying, also known as stop-and-go kinematic surveying, is not new. The original kinematic GPS innovator, Dr. Benjamin Remondi developed the idea in the mid-1980s. RTK is a method that can offer positional accuracy nearly as good as static carrier phase positioning, but RTK does it in real-time (Figure 9.2). Today, RTK has become routine in development and engineering surveys where the distance between the base and roving receivers can most often be measured in thousands of feet.

RTK is capable of delivering ±2 centimeter accuracy. Unlike DGPS that is differential GPS based on pseudoranges, RTK is a GPS method that uses carrier phase observations corrected in real-time and therefore, depends on the fixing of the integer cycle ambiguity.

Fixing the Integer Ambiguity in RTK

Most RTK systems resolve the integer ambiguity, *on-the-fly*. *On-the-fly* refers to a method of resolving the carrier phase ambiguity very quickly. The method requires dual-frequency GPS receivers capable of making both carrier phase and precise pseudorange measurements. The receiver is not required to remain stationary.

Here is one way it can be done. A search area is defined in the volume of the possible solutions, but that group is narrowed down quite a bit by using pseudoranges. If the number of integer combinations to be tested is greatly reduced with precise pseudoranges the search can be very fast. The possible solutions in that volume are tested statistically, according to a minimal variance criterion, and the best one is found. This candidate is verified, i.e., compared with the second best candidate.

The process can take less than 10 seconds under the best circumstances where the receivers are tracking a large constellation of satellites, the PDOP is small, the receivers are dual-frequency, there is no multipath, and the receiver noise is low.

This technique relies on dual-frequency information. Observations on L1 and L2 are combined into a widelane, which has an ambiguity of about 86 cm, and the integer ambiguity is solved in a first pass. This information is used to determine the kinematic solution on L1. Therefore, it is a good idea to restrict RTK to situations where there is good correlation of atmospheric biases at both ends of the baseline. In other words, RTK is best used when the distance between the base and rover is less than 20 km; this is usually not a problem.

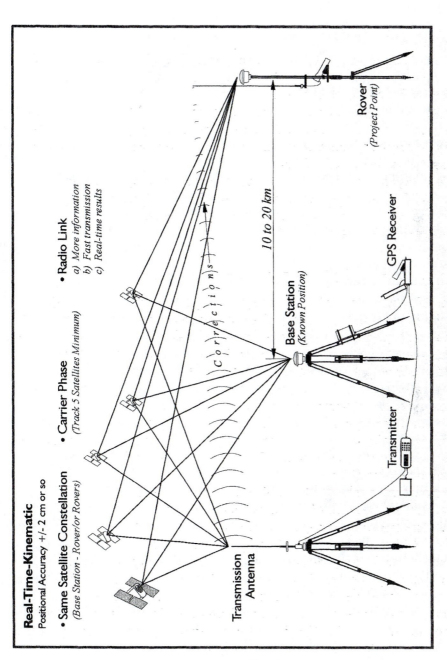

Figure 9.2. Real-Time Kinematic.

Radio License

RTK requires a radio link between the transmitter at the base station and the receiver at the rover, and they both must be tuned to the same frequency. The usual configuration operates at 4800 baud or faster. The units communicate with each other along a direct line-of-sight. In the United States most transmitters connected to RTK GPS surveying equipment operate between 450–470 MHz, and voice communications also operate in this same range.

The Federal Communications Commission (FCC) is concerned with some RTK GPS operations interfering with other radio signals, particularly voice communications. It is important for GPS surveyors to know that voice communications have priority over data communications.

The FCC requires cooperation among licensees that share frequencies. Interference should be minimized. For example, it is wise to avoid the most typical community voice repeater frequencies. They usually occur between 455-460 MHz and 465-470 MHz. Part 90 of the Code of Federal Regulations, 47 CFR 90, contains the complete text of the FCC rules.

There are also other international and national bodies that govern frequencies and authorize the use of signals elsewhere in the world. In some areas certain bands are designated for public use, and no special permission is required. For example, in Europe it is possible to use the 2.4 GHz band for spread-spectrum communication without special authorization, with certain power limitations. Here in the United States the band for spread-spectrum communication is 900 MHz.

Typical RTK

A typical RTK setup includes a base station and rover or rovers. They can be single- or dual-frequency receivers with GPS antennas, but dual-frequency receivers are usual. The radio receiving antennas for the rovers will either be built into the GPS antenna or they will be separate. At the base station the radio antenna for the data transmitter will most often be an omnidirectional whip antenna on a separate mast with a higher gain than those at the rovers.

It is usually best to place the transmitter antenna as high as is practical for maximum coverage. It is also best if the base station occupies a control station that has no overhead obstructions, is unlikely to be affected by multipath, and is somewhat away from the action if the work is on a construction site. It is also best if the base station is within line of sight of the rovers. If line of sight is not practical, as little obstruction as possible along the radio link is best.

A data radio transmitter consists of a radio modulator, an amplifier, and an antenna. The modulator converts the correction data into a radio signal. The amplifier increases the signal's power, which determines how far the information can travel. Well, not entirely; the terrain and the height of the antenna has something to do with it too. And then the information is transmitted via the antenna. RTK work requires a great deal of information be successfully communicated from the base station to the receivers. The base station transmitter ought to be

VHF, UHF, or spread-spectrum (frequency hopping or direct) to have sufficient capacity to handle the load. UHF and spread-spectrum radio modems are the most popular for DGPS and RTK applications. But while DGPS operations may need no more than 200 bps, bits-per-second, updated every 10 seconds or so, RTK requires at least 2,400 bps updated about every ½ second or less.

RTK is at its best when the distance between the base station and the rovers is less than 20 km, under most circumstances, but even before that limit is reached the radio data link can be troublesome. As stated earlier the Federal Communications Commission (FCC) is concerned with some RTK GPS operations interfering with other radio signals, particularly voice communications. Voice communications have priority over data communications.

In areas with high radio traffic it can be difficult to find an open channel. It is remarkable how often the interference emanates from other surveyors in the area doing RTK as well. That is why most radio data transmitters used in RTK allow the user several frequency options within the legal range. The radio link is vital for successful RTK.

It is also vital, of course, that the rover and the base station are tuned to the same frequency for successful communication. The receiver has an antenna and a demodulator. The demodulator converts the signal back to an intelligible form for the rover's receiver. The data signal from the base station can be weakened or lost at the rover from reflection, refraction, atmospheric anomalies, or even being too close. A rover that is too close to the transmitter may be overloaded and not receive the signal properly And, of course, even under the best circumstances the signal will fade as the distance between the transmitter and the rover grows too large.

The Vertical Component in RTK

The output of RTK can appear to be somewhat similar to that of optical surveying with an EDM and a level. The results can be immediate and with similar relative accuracy. Nevertheless, it is not a good idea to consider the methods equivalent. RTK offers some advantages, and some disadvantages when compared with more conventional methods. For example, RTK can be much more productive since it is available 24 hours a day and is not really affected by weather conditions. However, when it comes to the vertical component of surveying RTK and the level are certainly not equal.

GPS can be used to measure the differences in ellipsoidal height between points with good accuracy. However, unlike a level, unaided GPS cannot be used to measure differences in orthometric height. Or, as stated in Chapter 5, "orthometric elevations are not directly available from the geocentric position vectors derived from GPS measurements." The accuracy of orthometric heights in GPS is dependent on the veracity of the geoidal model used and the care with which it is applied.

Fortunately, ever improving geoid models have been, and still are, available from NGS. Since geoidal heights can be derived from these models, and ellipsoidal heights are available from GPS it is certainly feasible to calculate orthometric heights. In the past these calculations were done exclusively in postprocessed network adjustments. Today, more and more manufacturers are finding ways to include a geoid model in their RTK systems. However, it is important to remember that without a geoid model, RTK will only provide differences in ellipsoid heights between the base station and the rovers.

It is not a good idea to presume that the surface of the ellipsoid is sufficiently parallel to the surface of the geoid and ignore the deviation between the two. They may depart from one another as much as a meter in 4 or 5 kilometers. In other words, if the base station height is specified as an ellipsoidal height, then the heights at the rovers will be ellipsoidal. If the height at the base station is specified as an orthometric height then the heights at the rovers will be wrong, unless a geoid model is used in real-time by the rovers. Finally, it is certainly feasible to apply a geoid model to ellipsoidal heights in postprocessing with good results.

SOME PRACTICAL RTK SUGGESTIONS

Typical Satellite Constellations

In RTK, generally speaking, the more satellites that are available, the faster the integer ambiguities will be resolved. In the United States there are usually 6 satellites or so above an observer's horizon most of the time. And there are likely to be approximately 8 satellites above an observer's horizon about a third of the time; more only seldom. For baselines under 10 km an 8-satellite constellation should be quite adequate for good work under most circumstances.

Dual Frequency Receiver

A dual-frequency receiver is a real benefit in doing RTK. Using a dual-frequency receiver instead of a single-frequency receiver is almost as if there were one and a half more satellites available to the observer.

Setting up a Base Station

Set up the base station over a known position first, before configuring the rover. After the tripod and tribrach are level and over the point, attach the GPS antenna to the tribrach and, if possible, check the centering again.

Set up the base station transmitter in a sheltered location at least 10 feet from the GPS antenna, and close to the radio transmitter's antenna. It is best if the air flow of the base station transmitter's cooling fan is not restricted. The radio transmitter's antenna is often mounted on a range pole attached to a tripod. Set the radio transmitter's antenna as far as possible from obstructions and as high as stability will allow.

The base station transmitter's power is usually provided by a deep-cycle battery. Even though the cable is usually equipped with a fuse, it is best to be careful to not reverse the polarity when connecting it to the battery. It is also best to have the base station transmitter properly grounded, and avoid bending or kinking any cables.

After connecting the base station receiver to the GPS antenna, to the battery and the data collector, if necessary, carefully measure the GPS antenna height. This measurement is often the source of avoidable error, both at the base station and the rovers. Many surveyors measure the height of the GPS antenna to more than one place on the antenna, and it is often measured in both meters and feet for additional assurance.

Select a channel on the base station transmitter that is not in use, and be sure to note the channel used so that it may be set correctly on the rovers as well.

COMPARING RTK AND DGPS

Multipath in RTK and DGPS

While most other errors in GPS discussed earlier can be mediated or cancelled, due to the relatively short distance between receivers, the same cannot be said for multipath.

RTK, a method that is expected to collect high precision positions in real-time, is used to do stakeout and other location survey operations without postprocessing. And some environments that have the highest incidence of multipath, such as downtown city streets, are the places where RTK and DGPS are called upon most often. Yet in these techniques the position is computed over a very few epochs, there is little if any of the averaging of errors from epoch to epoch available in other applications. Therefore, in RTK the effect of multipath is rather direct, and even if it distorts the results only slightly, it may be too much in some situations.

In principle, multipath affects carrier observations the least. It is a bit worse for P-code pseudoranges. And in the observable used in most DGPS C/A-code pseudoranges, multipath affects the results the most. But receiver manufacturers have created ingenious technologies to minimize the multipath effects in DGPS pseudoranges. Therefore, it is entirely possible to find that it is in the RTK carrier phase application that multipath contributes the largest bias to the error budget. It can be reduced with choke-ring antennas, but they are somewhat heavy for easy field use.

Base Station

Concerning base stations in both RTK and DGPS, a base station on a coordinated control position must be available. Its observations must be simultaneous with those at the roving receivers and it must observe the same satellites.

It is certainly possible to perform a differential survey in which the position of the base station is either unknown, or based on an assumed coordinate, at the time of the survey. However, unless only relative coordinates are desired, the position of the base station must be known. In other words, the base station must occupy a control position, even if that control is established later.

Utilization of the DGPS techniques requires a minimum four satellites for three-dimensional positioning. RTK ought to have at least five satellites for initialization. Tracking five satellites is a bit of insurance against losing one abruptly; also it adds considerable strength to the results. While cycle slips are always a problem it is imperative in RTK that every epoch contain a minimum of four satellite data without cycle slips. This is another reason to always track at least five satellites when doing RTK. However, most receivers allow the number to drop to four after initialization.

Both methods most often rely on real-time communication between the base station and roving receivers. But RTK base station corrections are more complex than those required in DGPS.

For DGPS there are more sources of real-time correction signals all the time. There are commercial providers like *Omnistar* and *Landstar* that are satellite based. There are earthbound systems—*RDS* is an example. These sources offer code-phase corrections in the RTCM SC-104 format appropriate to DGPS. However, when it became clear in 1994 that including carrier phase information in the message could improve the accuracy of the system, RTCM Special Committee 104 added four new message types to Version 2.1 to fulfill the needs of RTK. Further improvements were made along the same line in Version 2.2 which became available in 1998. Nevertheless, proprietary message formats are perhaps more widely used in RTK work than in DGPS. In surveying applications, since RTK must work in a somewhat limited area it usually requires establishment of a project-specific base station.

Initialization

As mentioned, RTK relies on the carrier phase and the integer ambiguity must be solved. In other words, the method requires some time for initialization, usually at the start of day. Initialization is also required any time after which the continuous tracking of all available satellite signals stops for even the briefest length of time. After tracking of the same GPS satellites begins at the base station and the roving receiver or receivers, there is usually a short wait required for initialization to be accomplished. With many dual-frequency RTK receivers capable of reinitializing On-the-Fly, initialization can happen very quickly. Some receivers might require a return to a known coordinated position for reinitialization.

On the other hand, DGPS relies on pseudoranges; its immediate initialization is a clear advantage over some RTK systems. But DGPS can only provide meter accuracy, whereas RTK offers centimeter-level positional accuracies.

SUMMARY

So here are some of the fundamental ideas behind two methods. A fixed point with previously established coordinates is occupied by a base station. The base station computes the differences between its position and the positions measured at the roving receivers. The systematic errors in measurements are corrected at the roving receivers using the base station information. With DGPS the corrections can either be applied in postprocessing or in real-time.

And while these systematic errors can be virtually eliminated with differential correction, the biases such as multipath and receiver channel noise are certainly not eliminated. These errors have been much reduced in modern GPS receivers, but not completely defeated. Biases such as high PDOP can only be resolved with both good receiver design and care as to when and where the surveying is done.

RTK is still a fairly recent development. It is applicable to nearly any work that requires centimeter-level accuracy in real-time. It is used in engineering, surveying, air-navigation, mineral exploration, machine control, hydrography, and a myriad of other areas. It may be the latest GPS positioning technique, but it is certainly not the last. There will certainly be more efficient, accurate, faster, and easier GPS methods in the future.

EXERCISES

1. What is the meaning of latency as applied to DGPS and RTK?

   (a) the baud-rate of a radio modem in real-time GPS
   (b) the time taken for a system to compute corrections and transmit them to users in real-time GPS
   (c) the frequency of the RTCM SC104 correction signal in real-time GPS
   (d) the range rate broadcast with the corrections from the base station

2. Which of the following errors are not reduced by using DGPS or RTK methods?

   (a) atmospheric errors
   (b) satellite clock bias
   (c) ephemeris bias
   (d) multipath

3. Which of the following statements about RTCM SC104 is not true?

   (a) RTCM SC104 is a format for communication used in real-time GPS.
   (b) RTCM SC104 was originally designed in 1985.
   (c) RTCM SC104 can be used in both DGPS and RTK applications.
   (d) RTCM SC104 was revised in 1994 and 1996.

4. Who was the original kinematic GPS innovator?

   (a) Ann Bailey
   (b) Benjamin Remondi
   (c) George F. Syme
   (d) R.E. Kalman

5. Which of the following statements about rules governing RTK radio communication is correct?

   (a) According to the FCC regulations voice communications have priority over data communications.
   (b) Code of Federal Regulations, 48 CFR 91, contains the complete text of the FCC Rules.
   (c) Most typical community voice repeater frequencies are about 900 MHz.
   (d) In Europe it is possible to use the 4.2 GHz band for spread-spectrum communication without special authorization.

6. Which of the following data rates is closest to that required for RTK?

   (a) 200 bps updated every 10 seconds
   (b) 1,400 bps updated about every 5 seconds
   (c) 2,400 bps updated about every ½ of a second
   (d) 3,500 bps updated about every 0.05 of a second

7. How many satellites is the minimum for the best initialization with RTK?

   (a) 4 satellites
   (b) 5 satellites
   (c) 6 satellites
   (d) 8 satellites

8. What is a typical range of systematic error that can currently be expected in an autonomous point position?

   (a) ±100 m to ±120 m
   (b) ±75 m to ±95 m
   (c) ±45 m to ±65 m
   (d) ±20 m to ±40 m

9. Which of the following is not a part of a typical data radio transmitter used in RTK?

   (a) a radio modulator
   (b) an amplifier
   (c) a demodulator
   (d) an antenna

10. Which of the following does not correctly describe a difference between RTK and DGPS?

    (a) RTK relies on the carrier phase and the integer ambiguity must be solved, but with DGPS it does not.
    (b) The RTK base station corrections are more complex than those required in DGPS.

(c) RTK offers centimeter-level positional accuracies, DGPS provides meter-level positional accuracies.

(d) RTK is affected by multipath, but DGPS is not.

ANSWERS AND EXPLANATIONS

1. Answer is (b)

Explanation: In DGPS and RTK it takes some time for the base station to calculate corrections and it takes some time for it to transmit them. The base station's data are put into packets in the correct format. The data makes its way from the base station to the rover over the data link. It must then be decoded and go through the rover's software. The time it takes for all of this to happen is called the *latency* of the communication between the base station and the rover. It can be as little as a quarter of a second or as long as a couple of seconds.

2. Answer is (d)

Explanation: In autonomous point positioning all the biases discussed earlier affect the GPS position. For example, in autonomous point positioning the satellite clock bias normally contributes about 3 m to the final GPS positional error, ephemeris bias contributes about 5 m, and solar radiation, etc., contribute about another 1.5 m. But with DGPS and RTK, when errors are well correlated all these biases are virtually zero, as are the atmospheric errors, unlike autonomous point positioning where the ionospheric delay contributes about 5 m and the troposphere about 1.5 m. While most other errors in GPS discussed earlier can be mediated, or cancelled, due to the relatively short distance between receivers, the same cannot be said for multipath.

3. Answer is (d)

Explanation: In DGPS there is an agreed-upon format for the communication between the base station and rovers. It was first designed for marine navigation. In 1985, *the Radio Technical Commission for Maritime Services, RTCM, Study Committee, SC-104,* created a standard format for DGPS corrections that is still more used than any proprietary formats that have come along since. Both the United States and the Canadian Coast Guards use this format in their coastal beacons. When it became clear in 1994 that including carrier phase information in the message could improve the accuracy of the system, RTCM Special Committee 104 added four new message types to Version 2.1 to fulfill the needs of RTK. Further improvements were made along the same line in Version 2.2 which became available in 1998.

4. Answer is (b)

   Explanation: Kinematic surveying, also known as stop-and-go kinematic surveying, is not new. The original kinematic GPS innovator, Dr. Benjamin Remondi developed the idea in the mid-1980s.

5. Answer is (a)

   Explanation: In the United Stated most transmitters connected to RTK GPS surveying equipment operate between 450–470 MHz, and voice communications also operate in this same range. It is important for GPS surveyors to know that voice communications have priority over data communications. The FCC requires cooperation among licensees that share frequencies. Interference should be minimized. For example, it is wise to avoid the most typical community voice repeater frequencies. They usually occur between 455-460 MHz and 465–470 MHz. Part 90 of the Code of Federal Regulations, 47 CFR 90, contains the complete text of the FCC Rules.

   There are also other international and national bodies that govern frequencies and authorize the use of signals elsewhere in the world. In some areas certain bands are designated for public use, and no special permission is required. For example, in Europe it is possible to use the 2.4 GHz band for spread-spectrum communication without special authorization with certain power limitations. Here in the United States the band for spread-spectrum communication is 900 MHz.

6. Answer is (b)

   Explanation: DGPS operations may need no more than 200 bps, bits-per-second, updated every 10 seconds or so; RTK requires at least 2,400 bps updated about every ½ second or less.

7. Answer is (b)

   Explanation: Utilization of the DGPS techniques requires a minimum four satellites for three-dimensional positioning. RTK ought to have at least five satellites for initialization. Tracking five satellites is a bit of insurance against losing one abruptly; also it adds considerable strength to the results. While cycle slips are always a problem it is imperative in RTK that every epoch contain a minimum of four satellite data without cycle slips. This is another reason to always track at least five satellites when doing RTK. However, most receivers allow the number to drop to four after initialization.

8. Answer is (d)

   Explanation: An autonomous point position in GPS with SA turned off will usually have about ± 20 m to ± 40 m of systematic bias in a horizontal position, whereas a DGPS position may be subject to a tenth of that, or even less.

9. Answer is (c)

    Explanation: A data radio transmitter consists of a radio modulator, an ampli-fier, and an antenna. The modulator converts the correction data into a radio signal. The amplifier increases the signal's power. And then the information is transmitted via the antenna.

10. Answer is (d)

    Explanation: Both methods most often rely on real-time communication be-tween the base station and roving receivers. But RTK base station correc-tions are more complex than those required in DGPS. In surveying applications, since RTK must work in a somewhat limited area it usually requires estab-lishment of a project-specific base station. RTK relies on the carrier phase and the integer ambiguity must be solved. On the other hand, DGPS relies on pseudoranges; its immediate initialization is a clear advantage over some RTK systems. But DGPS can only provide meter accuracy, whereas RTK offers centimeter-level positional accuracies. Both techniques are affected by multipath.

Glossary

Absolute Positioning
(aka Point Positioning, Point Solution, or Single Receiver Positioning)
A single receiver position, defined by a coordinate system. Most often the coordinate system used in absolute positioning is geocentric. In other words, the origin of the coordinate system is intended to coincide with the center of mass of the earth.

Accuracy
The agreement of a value, whether measured or computed, with the standard or accepted true value. In the absolute sense, the true value is unknown and therefore, accuracy can only be estimated. Nevertheless, in measurement, accuracy is considered to be directly proportional to the attention given to the removal of systematic errors and mistakes. In GPS specifically, the values derived are usually the position, time, or velocity at GPS receivers.

Almanac
A data file containing a summary of the orbital parameters of all GPS satellites. The almanac is found in subframe 5 of the Navigation Message. This information can be acquired by a GPS receiver from a single GPS satellite; it helps the receiver find the other GPS satellites.

Analog
Representation of data by a continuous physical variable. For example, the modulated carrier wave used to convey information from a GPS satellite to a GPS receiver is an analog mechanism.

Antenna
An antenna is a resonant device that collects and often amplifies a satellite's signals. It converts the faint GPS signal's electromagnetic waves into electric currents sensible to a GPS receiver. Microstrip, also known as patch, antennas are the most often used with GPS receivers. Choke-ring antennas are intended to minimize multipath error.

Antenna Splitter
(see Zero Baseline)
An antenna splitter is an attachment which divides a single GPS antenna's signal in two. The divided signal goes to two GPS receivers. This is the foundation

of Zero Baseline test. The two receivers are using the same antenna so the base line length should be measured as zero. Since perfection remains elusive, it is usually a bit more.

Antispoofing (AS)

The encryption of the P code to render spoofing ineffective. Spoofing is generation of false transmissions masquerading as the Precise Code (P Code). This countermeasure, Antispoofing, is actually accomplished by the modulation of a W Code to generate the more secure Y Code that replaces the P code. Commercial GPS receiver manufacturers are not authorized to use the P code directly. Therefore, most have developed proprietary techniques both for carrier wave and pseudorange measurements on L2 indirectly. Dual-frequency GPS receivers must also overcome AS. Antispoofing was first activated on all Block II satellites on January 31, 1994.

Anywhere Fix

A receiver's ability to achieve lock-on without being given a somewhat correct beginning position and time.

Atomic Clock

(see Cesium Clock)

A clock regulated by the resonance frequency of atoms or molecules. In GPS satellites the substances used to regulate atomic clocks are cesium, hydrogen, or rubidium.

Attosecond

One quintillionth (10^{18}) of a second.

Attribute

Information about features of interest. Attributes such as date, size, material, color, etc., are frequently recorded during data collection for Geographic Information Systems (GIS).

Availability

The period, expressed as a percentage, when positioning from a particular system is likely to be successful.

Bandwidth

The bandwidth of a signal is its range of frequencies. It is a measurement of the difference between the highest frequency and the lowest frequency expressed in Hertz. For example, the bandwidth of a voice signal is about 3 kHz, but for television signals it is around 6 MHz.

Baseline

A line described by two stations from which GPS observations have been made simultaneously. A vector of coordinate differences between this pair of stations is one way to express a baseline.

Base Station

(aka Reference Station)

A known location where a static GPS receiver is set. The base station is intended to provide data with which to differentially correct the GPS measurements collected by one or more roving receivers. A base station used in Differential GPS (DGPS) is used to correct pseudorange measurements collected by roving receivers to improve their accuracy. In carrier phase surveys the base station data are combined with the other receivers' measurements to calculate double-differenced observations or used in real-time kinematic (RTK) configurations.

Beat Frequency

When two signals of different frequencies are combined, two additional frequencies are created. One is the sum of the two original frequencies. One is the difference of the two original frequencies. Either of these new frequencies can be called a beat frequency.

Between-Epoch Difference

The difference in the phase of the signal on one frequency from one satellite as measured between two epochs observed by one receiver.

Because a GPS satellite and a GPS receiver are always in motion relative to one another the frequency of the signal broadcast by the satellite is not the same as the frequency received. Therefore, the fundamental Doppler observable in GPS is the measurement of the change of phase between two epochs.

Between-Receiver Single Difference

The difference in the phase measurement between two receivers simultaneously observing the signal from one satellite on one frequency.

For a pair of receivers simultaneously observing the same satellite, a between-receiver single difference pseudorange or carrier phase observable can virtually eliminate errors attributable to the satellite's clock. When baselines are short the between-receiver single difference can also greatly reduce errors attributable to orbit and atmospheric discontinuities.

Between-Satellite Single Difference

The difference in the phase measurement between signals from two satellites on one frequency simultaneously observed by one receiver.

For one receiver simultaneously observing two satellites, a between-satellite single difference pseudorange or carrier phase observable can virtually eliminate errors attributable to the receiver's clock.

Bias

A systematic error.

Biases affect all GPS measurements, and hence the coordinates and baselines derived from them. Biases may have a physical origin, satellite orbits, atmospheric conditions, clock errors, etc. They may also originate from less than perfect control coordinates, incorrect ephemeris information, etc. Modeling is one method used to eliminate, or at least limit the effect of biases.

Binary Biphase Modulation

The method used to impress the pseudorandom noise codes onto GPS carrier waves using two states of phase modulations, since the code is binary. In binary biphase modulation, the phase changes that occur on the carrier wave are either 0° or 180°.

Bit

A unit of information. In a binary system, either a 1 or a 0.

Block I, II, IIR, IIF Satellites

Classification of the GPS satellite's generations.

Block I satellites were officially experimental. There were 11 Block I satellites. PRN4, the first GPS satellite, was launched February 22, 1978. The last Block I satellite was launched October 9, 1985, and none remain in operation.

Block II satellites are operational satellites. The GPS constellation was declared operational in 1995. Twenty-seven Block II satellites have been launched.

Block IIR are replenishment satellites and have the ability to receive signals from other IIR satellites. These satellites have the capability to measure ranges between themselves.

Block IIF will be the follow-on series of satellites in the new century.

Broadcast

A modulated electromagnetic wave transmitted across a large geographical area.

Broadcast Ephemeris

(see Ephemeris)

Byte

A sequence of eight binary digits that represents a single character; i.e., a letter, a number.

C/A code

(aka S code)

A binary code known by the names Civilian/Access code, Clear/Access code, Clear/Acquisition code, Coarse/Acquisition code and various other combinations of similar words. It is a standard-spread spectrum GPS pseudorandom noise code modulated on the L1 carrier using binary biphase modulations. The C/A code is not carried on L2. Each C/A code is unique to the particular GPS satellite broadcasting it. It has a chipping rate 1.023 MHz, more than a million bits per second. The C/A code is a direct sequence code and a source of information for pseudorange measurements for commercial GPS receivers.

Carrier

An electromagnetic wave, usually sinusoidal, that can be modulated to carry information. Common methods of modulation are frequency modulation and amplitude modulation. However, in GPS the phase of the carrier is modulated and there are currently two carrier waves. The two carriers in GPS are L1 and L2, which are broadcast at 1575.42 MHz and 1227.60 MHz, respectively.

Carrier Beat Phase

(see Beat Frequency)
(also see Doppler Shift)

The phase of a beat frequency created when the carrier frequency generated by a GPS receiver combines with an incoming carrier from a GPS satellite. The two carriers have slightly different frequencies as a consequence of the Doppler Shift of the satellite carrier compared with the nominally constant receiver-generated carrier. Because the two signals have different frequencies, two beat frequencies are created. One beat is the sum of the two frequencies. One is the difference of the two frequencies.

Carrier Frequency

In GPS the frequency of the unmodulated signals broadcast by the satellites. Currently, the carrier frequencies are L1 and L2, which are broadcast at 1575.42 MHz and 1227.60 MHz, respectively.

Carrier Phase

(aka Reconstructed Carrier Phase)

(1) The term is usually used to mean GPS measurements based on the carrier signal itself, either L1 or L2, rather than measurements based on the codes modulated onto the carrier wave.

(2) Carrier phase may also be used to mean a part of a full carrier wavelength. An L1 wavelength is about 19 cm, an L2 wavelength is about 24 cm. A part of the full wavelength may be expressed in a phase angle from 0° to 360°, a fraction of a wavelength, cycle, or some other way.

Carrier Tracking Loop

(see also Code Tracking Loop)

A feedback loop that a GPS receiver uses to generate and match the incoming carrier wave from a GPS satellite.

CEP

(see Circular Error Probable)

Cesium Clock

(aka Cesium Frequency Standard)

An atomic clock that is regulated by the element cesium. Cesium atoms, specifically atoms of the isotope Cs-133, are exposed to microwaves and the atoms vibrate at one of their resonant frequencies. By counting the corresponding cycles time is measured.

CGSIC
Civil GPS Service Interface Committee.

Channel
A channel of a GPS receiver consists of the circuitry necessary to receive the signal from a single GPS satellite on one of the two carrier frequencies. A channel includes the digital hardware and software.

Checkpoint
A reference point from an independent source of higher accuracy used in the estimation of the positional accuracy of a data set.

Chipping Rate
In GPS the rate at which chips, binary 1s and 0s, are produced. The P code chipping rate is about 10 million bits per second. The C/A code chipping rate is about 1 million bits per second.

Circular Error Probable (CEP)
A description of two-dimensional precision. The radius of a circle, with its center at the actual position, that is expected to be large enough to include half (50%) of the normal distribution of the scatter of points observed for the position.

Civilian Code
(see C/A code)

Civilian/Access Code
(see C/A code)

Class of Survey
Class of Survey is a means to generalize and prioritize the precision of surveys. The foundation of the categories are often taken from geodetic surveying classifications supplemented with standards of higher precision applicable to GPS work. Different classifications frequently apply to horizontal and vertical surveys. The categories themselves, their notation, and accuracy tolerances are unique to each nation. A Class of Survey should reflect the quality of network design, the instruments, and methods used and processing techniques as well as the precision of the measurements.

Clear/Access Code
(see C/A code)

Clear/Acquisition Code
(see C/A code)

Clock Bias

The discrepancy between a moment of time per a GPS receiver's clock or a GPS satellite's clock and the same moment of time per GPS Time, or another reference, such as Coordinated Universal Time or International Atomic Time.

Clock Offset

A difference between the same moment of time as indicated by two clocks.

Coarse Acquisition (C/A)

(see C/A code)

Code Phase

Measurements based on the C/A code rather than measurements based on the carrier waves. Sometimes used to mean pseudorange measurements expressed in units of cycles.

Code Tracking Loop

A feedback loop used by a GPS receiver to generate and match the incoming codes, C/A or P codes, from a GPS satellite.

Complete Instantaneous Phase Measurement

(see Fractional Instantaneous Phase)

A measurement including the integer number of cycles of carrier beat phase since the initial phase measurement.

Confidence Level

The probability that the true value is within a particular range of values, expressed as a percentage.

Confidence Region

A region within which the true value is expected to fall, attended by a confidence level.

Constellation

(1) The Space Segment, all GPS satellites in orbit.
(2) The particular group of satellites used to derive a position.
(3) The satellites available to a GPS receiver at a particular moment.

Continuously Operating Reference Stations (CORS)

A system of base stations that originated from the stations built to support air and marine navigation with real-time differential GPS correction signals.

Continuous-Tracking Receiver

(see Multichannel Receiver)

Control Segment

The U.S. Department of Defense's network of GPS monitoring and upload stations around the world. The tracking data derived from this network is used to prepare, among other things, the broadcast ephemerides and clock corrections by the Master Control Station at Schriever Air Force Base, Colorado Springs, Colorado. Then included in the Navigation Message, the calculated information is uploaded to the satellites.

Coordinated Universal Time
(aka UTC, Zulu Time)

Universal Time systems are international, based on atomic clocks around the world. Coordinated atomic clock time called Tempes Atomique International (TAI) was established in the 1970s and it is very stable. It is more stable than the actual rotation of the earth. TAI would drift out of alignment with the planet if leap seconds were not introduced periodically as they are in UTC. UTC is in fact one of several of the Universal Time standards. There are standards more refined than UTC's tenth of a second. UTC is maintained by the U.S. Naval Observatory (USNO).

Correlation-Type Channel

A channel in a GPS receiver used to shift or compare the incoming signal with an internally generated signal. The code generated by the receiver is cross-correlated with the incoming signal to find the correct delay. Once they are aligned, the delay lock loop keeps them so. Correlator designs are sometimes optimized for acquisition of signal under foliage, accuracy, multipath mitigation, etc.

Crosstrack Error

The distance from a particular trajectory to the desired course.

Cutoff Angle
(see Mask Angle)

Cycle Ambiguity
(aka Integer Ambiguity)

The number of full wavelengths, the integer number of wavelengths, between a particular receiver and satellite is initially unknown in a carrier phase measurement. It is called the cycle or integer ambiguity. If a single frequency receiver is tracking several satellites there is a different ambiguity for each. If a dual-frequency receiver is tracking several satellites, there is a different ambiguity for each satellite's L1 and L2. However, in every case the ambiguity is a constant number as long as the tracking, lock, is not interrupted. However, should the signal be blocked, if a cycle slip occurs, there is a new ambiguity to be resolved.

Once the initial integer ambiguity value is resolved in a fixed solution for each satellite-receiver pair the integrated carrier phase measurements can yield very precise positions. But the cycle ambiguity resolution processes actually utilize double-differenced carrier phase observables, not single satellite-receiver measurements. The great progress in carrier phase GPS systems has reduced the length of observation data needed, as in rapid static techniques. The integer ambiguity resolution can now occur even with the receiver in motion, the on-the-fly approach.

Cycle Slip
A discontinuity of an integer number of cycles in the carrier phase observable. A jump of a whole number of cycles in the carrier tracking loop of a GPS receiver. Usually the result of a temporarily blocked GPS signal.

A cycle slip causes the cycle ambiguity to change suddenly. The repair of a cycle slip includes the discovery of the number of missing cycles during an outage. Cycle slips must be repaired before carrier phase data can be successfully processed in double-differenced observables.

Datalogger
(aka Data Recorder and Data Collector)
A data entry computer, usually small, lightweight and often handheld.

A datalogger stores information supplemental to the measurements of a GPS receiver.

Data Message
(see Navigation Message)

Data Set
An organized collection of related data compiled specifically for computer processing.

Data Transfer
Transporting data from one computer or software to another, often accompanied by a change in the format of the data.

Datum
(1) Any reference point line or surface used as a basis for calculation or measurement of other quantities.

(2) A means of relating coordinates determined by any means to a well-defined reference frame.

(3) The singular of data.

Decibel
(abbreviated dB)
Decibel, a tenth of a bel. The bel was named for Alexander Graham Bell. Decibel does not actually indicate the power of the antenna, it refers to a comparison. Most GPS antennas have a gain of about 3 dB, decibels. This indicates that the GPS antenna has about 50 % of the capability of an *isotropic* antenna. An isotropic antenna is a hypothetical, lossless antenna that has equal capabilities in all directions.

DGPS
(see Differential GPS)

Differential GPS
(1) A method that improves GPS pseudorange accuracy. A GPS receiver at a base station, a known position, measures pseudoranges to the same satellites at the same time as other roving GPS receivers. The roving receivers occupy unknown positions in the same geographic area. Occupying a known position, the base station receiver finds corrective factors that can either be communicated in real-time to the roving receivers, or may be applied in postprocessing.

(2) The term Differential GPS is sometimes used to describe Relative Positioning. In this context it refers to more precise carrier phase measurement to determine the relative positions of two receivers tracking the same GPS signals in contrast to Absolute, or Point, Positioning.

Digital
Involving or using numerical digits. Information in a binary state, either a one or a zero, a plus or a minus, is digital. Computers utilize the digital form almost universally.

Dilution of Precision (DOP)
In GPS positioning, an indication of geometric strength of the configuration of satellites in a particular constellation at a particular moment and hence the quality of the results that can be expected. A high DOP anticipates poorer results than a low DOP. A low DOP indicates that the satellites are widely separated. Since it is based solely on the geometry of the satellites, DOP can be computed without actual measurements.

There are various categories of DOP, depending on the components of the position fix that are of most interest.

Standard varieties of Dilution of Precision:

PDOP: Position (three coordinates)
GDOP: Geometric (three coordinates and clock offset)
RDOP: Relative (normalized to 60 seconds)
HDOP: Horizontal (two horizontal coordinates)
VDOP: Vertical (height)
TDOP: Time (clock offset)

Distance Root Mean Square (drms)
A statistical measurement that can characterize the scatter in a set of randomly varying measurements on a plane.

The drms is calculated from a set of data by finding the root-mean-square value of the radial errors from the mean position. In other words, the root-mean-square value of the linear distances between each measured position and the ostensibly true location.

In GPS positioning, 2 drms is more commonly used. 2 drms does *not* mean, two-dimensional (2D) rms. 2 drms *does* mean twice the distance root mean square. In practical terms, a particular 2 drms value is the radius of a circle that is expected to contain from 95 to 98% of the positions a receiver collects in one occupation, depending on the nature of the particular error ellipse involved. 2 drms is convenient to calculate. It can be predicted using covariance analysis by multiplying the HDOP by the standard deviation of the observed pseudoranges.

Dithering
Intentional introduction of digital noise. Dithering the satellite's clocks is the method the Department of Defense (DOD) uses to degrade the accuracy of the Standard Positioning Service. This degradation was known as Selective availability. Selective availability was switched off on May 2, 2000 by presidential order.

DOD
Department of Defense

DOP
(see Dilution of Precision)

Doppler-Aiding
A method of receiver signal processing that relies on the Doppler Shift to smooth tracking.

Doppler Shift
In GPS, a systematic change in the apparent frequency of the received signal caused by the motion of the satellite and receiver relative to one other.

DOT
Department of Transportation

Double-Difference
A method of GPS data processing. In this method simultaneous measurements made by two GPS receivers, either pseudorange or carrier phase, are combined. Both satellite and receiver clock errors are virtually eliminated from the solution. Most high precision GPS positioning methods use double-difference processing in some form.

Dual-Capability GPS/Glonass Receiver
(aka Interoperability)
A receiver that has the capability to track both GPS satellites and GLONASS satellites.

Dual-Frequency
Receivers or GPS measurements that utilize both L1 and L2. Dual-frequency implies that advantage is taken of pseudorange and/or carrier phase on both L-band

frequencies. Dual-frequency allows modeling of ionospheric bias and attendant improvement in long baseline measurements particularly.

Dynamic Positioning
(see Kinematic Positioning)

Earth-Centered Earth-Fixed
A Cartesian system of coordinates with three axes in which the origin of the 3D system is the Earth's center of mass, the geocenter. The z-axis passes through the North Pole, that is the International Reference Pole (IRP) as defined by the International Earth Rotation Service (IERS). The x-axis passes through the intersection of zero longitude, near the Greenwich meridian, and the equator. The y-axis extends through the geocenter along a line perpendicular from the x-axis. It completes the right-handed system with the x- and z-axes and all three rotate with the earth. Three coordinate reference systems—NAD 83, WGS 84, and ITRS—are all similarly defined.

ECEF
(see Earth-Centered Earth-Fixed)

Elevation
The distance measured along the direction of gravity above a surface of constant potential. Usually the reference surface is the geoid. Mean sea level (MSL) once utilized as a reference surface approximating the geoid is known to differ from the geoid up to a meter or more.

The term height, sometimes considered synonymous with elevation, refers to the distance above an ellipsoid in geodesy.

Elevation Mask Angle
(see Mask Angle)

Ellipsoid of Revolution
(aka spheroid)

A biaxial closed surface, whose planar sections are either ellipses or circles, that is formed by revolving an ellipse about its minor axis. Two quantities fully define an ellipsoid of revolution, the semimajor axis, a, and the flattening, $f = (a - b)/a$, where b is the length of the semiminor axis.

The computations for the North American Datum of 1983 were done with respect to the GRS80 ellipsoid, the Geodetic Reference System of 1980. The GRS80 ellipsoid is defined by:

$$a = 6738137 \text{ meters}$$

$$1/f = 298.257222101$$

Ellipsoidal Height
(aka Geodetic Height)
The distance from an ellipsoid of reference to a point on the Earth's surface, as measured along the perpendicular from the ellipsoid. Ellipsoidal height is not the same as elevation above mean sea level nor orthometric height. Ellipsoidal height also differs from geoidal height.

Ephemeris
A table of values including locations and related data from which it is possible to derive a satellite's position and velocity. In GPS, broadcast ephemerides are compiled by the Master Control Station, uploaded to the satellites by the Control Segment, and transmitted to receivers in the Navigation Message. It is designed to provide orbital elements quickly and is not as accurate as the precise ephemeris. Broadcast ephemeris errors are mitigated by differential correction or in double-differenced observables over short baselines. Precise ephemerides are postprocessed tables available to users via the Internet.

Epoch
A time interval. In GPS, the period of each observation in seconds.

Error Budget
A summary of the magnitudes and sources of statistical errors that can help in approximating the actual errors that will accrue when observations are made.

FAA
Federal Aviation Administration

Fast-Switching Channel
(see Multiplexing Channel)

Federal Radionavigation Plan
A document mandated by Congress that is published every other year by the Department of Defense and the Department of Transportation. In an effort to reduce the functional overlap of federal initiatives in radionavigation, it summarizes plans for promotion, maintenance, and discontinuation of domestic and international systems.

Femtosecond
One millionth of a nanosecond. It is 10^{15} of a second.

FM
Frequency Modulation

Fractional Instantaneous Phase
A carrier beat phase measurement not including the integer cycle count. It is always between zero and one cycle.

Frequency
(see Wavelength)
The number of cycles per unit of time. In GPS, the frequency of the unmodulated carrier waves are L1 at 1575.42 MHz, and L2 at 1227.60 MHz.

Frequency Band
Within the electromagnetic spectrum, a particular range of frequencies.

FRP
Federal Radionavigation Plan

Gain
The *gain*, or *gain pattern* of a GPS antenna refers to its ability to collect more energy from above the mask angle, and less from below the mask angle.

General Theory of Relativity
(see also Special Theory of Relativity)
A physical theory published by A. Einstein in 1916. In this theory space and time are no longer viewed as separate but rather form a four-dimensional continuum called space-time that is curved in the neighborhood of massive objects. The theory of general relativity replaces the concept of absolute motion with that of relative motion between two systems or frames of reference and defines the changes that occur in length, mass, and time when a moving object or light pass through a gravitational field.

General Relativity predicts that as gravity weakens, the rate of clocks increase—they tick faster. On the other hand, Special Relativity predicts that moving clocks appear to tick more slowly than stationary clocks, since the rate of a moving clock seems to decrease as its velocity increases. Therefore, for GPS satellites, General Relativity predicts that the atomic clocks in orbit on GPS satellites tick faster than the atomic clocks on earth by about 45,900 ns/day. Special Relativity predicts that the velocity of atomic clocks moving at GPS orbital speeds tick slower by about 7,200 ns/day than clocks on earth. The rates of the clocks in GPS satellites are reset before launch to compensate for these predicted effects.

Geodesy
The science concerned with the size and shape of the earth. Geodesy also involves the determination of positions on the earth's surface and the description of variations of the planet's gravity field.

Geodetic Datum
A model defined by an ellipsoid and the relationship between the ellipsoid and the surface of the earth, including a Cartesian coordinate system. In modern usage, eight constants are used to form the coordinate system used for geodetic control. Two constants are required to define the dimensions of the reference ellipsoid.

Three constants are needed to specify the location of the origin of the coordinate system, and three more constants are needed specify the orientation of the coordinate system. In the past, a geodetic datum was defined by five quantities: the latitude and longitude of an initial point, the azimuth of a line from this point, and the two constants to define the reference ellipsoid.

Geodetic Survey

A survey that takes into account the size and shape of the Earth. A geodetic survey may be performed using terrestrial or satellite positioning techniques.

Geographic Information System (GIS)

A computer system used to acquire, store, manipulate, analyze, and display spatial data. A GIS can also be used to conduct analysis and display the results of queries.

Geoid

The equipotential surface of the Earth's gravity field which approximates mean sea level most closely. The geoid surface is everywhere perpendicular to gravity. Several sources have contributed to models of the Geoid, including ocean gravity anomalies derived from satellite altimetry, satellite-derived potential models, and land surface gravity observations.

Geoidal Height

The distance from the ellipsoid of reference to the geoid measured along a perpendicular to the ellipsoid of reference.

Geometrical Dilution of Precision (GDOP)

(see Dilution of Precision)

Gigahertz (GHz)

One billion cycles per second.

Global Navigation Satellite System (GNUS)

A satellite-based positioning system. In Europe GNUS-1 is a reference to a combination of GPS and GLONASS. However, GNUS-2 is a reference to a proposed combination of GPS, GLONASS, and other systems both space- and ground-based.

Global Orbiting Navigation Satellite System (GLONASS)

A satellite radionavigation system financed by the Soviet Commonwealth. It is comprised of 21 satellites and 3 active spares arranged in three orbital rings approximately 11,232 nautical miles above the earth, though the number of functioning satellites may vary because of funding. Frequencies are broadcast in the ranges of 1597–1617 MHz and 1240–1260 MHz. GLONASS positions reference the datum PZ90 rather than WGS84.

Global Positioning System (GPS)

A radionavigation system for providing the location of GPS receivers with great accuracy. Receivers may be stationary on the surface of the Earth or in vehicles: aircraft, ships, or in Earth-orbiting satellites. The Global Positioning System is comprised of three segments; the User Segment, the Space Segment, and the Control Segment.

GPSIC

GPS Information Center, United States Coast Guard.

GPS Receiver

An apparatus that captures modulated GPS satellite signals to derive measurements of time, position, and velocity.

GPS Time

GPS time is the time given by all GPS Monitoring Stations and satellite clocks. GPS time is regulated by Coordinated Universal Time (UTC). GPS time and Coordinated Universal Time were the same at midnight UT on January 6, 1980. Since then GPS time has not been adjusted as leap seconds were inserted into UTC approximately every 18 months. These leap seconds keep UTC approximately synchronized with the Earth's rotation. GPS time has no leap seconds and is offset from UTC by an integer number of seconds; the number of seconds is in the Navigation Message and most GPS receivers apply the correction automatically. The exact difference is contained in two constants, within the Navigation Message, A0 and A1, providing both the time difference and rate of system time relative to UTC. Disregarding the leap second offset, GPS time is mandated to stay within one microsecond of UTC, but over several years it has actually remained within a few hundred nanoseconds.

GRS80

Geocentric Reference System 1980. An ellipsoid adopted by the International Association of Geodesy.

Major semiaxis (a) = 6378137 meters
Flattening (1/f) = 298.257222101

Ground Antennas

The S-band antennas used to upload information to the GPS satellites, including broadcast ephemeris data and broadcast clock corrections.

Handover Word

Information used to transfer tracking from the C/A code to the P code. This time synchronization information is in the Navigation Message.

Height, Ellipsoidal
(see Ellipsoidal Height)

Height, Orthometric
(see Orthometric Height)

Hertz
One cycle per second.

Heuristic Approach
Fancy words for trial and error.

Hydrogen Maser
An atomic clock. A device that uses microwave amplification by stimulated emission of radiation is called a maser. Actually, the microwave designation is not entirely accurate since masers have been developed to operate at many wavelengths. In any case, a maser is an oscillator whose frequency is derived from atomic resonance. One of the most useful types of maser is based on transitions in hydrogen, which occur at 1,421 MHz. The hydrogen maser provides a very sharp, constant oscillating signal, and thus serves as a time standard for an atomic clock.

The active hydrogen maser provides the best-known frequency stability for a frequency generator commercially available. At a 1 hour averaging time the active maser exceeds the stability of the best-known cesium oscillators by a factor of at least 100, and the hydrogen maser is also extremely environmentally rugged.

Independent Baselines
(aka Nontrivial baselines)
These are baselines observed using GPS Relative Positioning techniques. When more than two receivers are observing at the same time, both independent (nontrivial) and trivial baselines are generated. For example, where r is the number of receivers, every complete static session yields r-1 independent (nontrivial) baselines. If four receivers are used simultaneously, six baselines are created. However, three of those lines will fully define the position of each occupied station in relation to the others. Therefore, the user can consider three (r-1) of the six lines independent (nontrivial), but once the decision is made only those three baselines are included in the network.

Inmarsat
International Maritime Satellite.

Integer Ambiguity
(see Cycle Ambiguity)

Integrity
A quality measure of GPS performance including a system to provide a warning when the system should not be used for navigation because of some inadequacy.

Interferometry
(see Relative Positioning)

International Earth Rotation Service (IERS)
   IERS was created in 1988 by the International Union of Geodesy and Geophysics (IUGG) and the International Astronomical Union (IAU). It is an interdisciplinary service organization that includes astronomy, geodesy, and geophysics. IERS maintains the International Celestial Reference System, and the International Terrestrial Reference System.

International GPS Service (IGS)
   The IGS is comprised of many civilian agencies that operate a worldwide GPS tracking network. IGS produces postmission ephemerides, tracking station coordinates, earth orientation parameters, satellite clock corrections, tropospheric and ionospheric models. It is an initiative of the International Association of Geodesy, and other scientific organizations, established in 1994 and originally intended to serve precise surveys for monitoring crustal motion. The range of users has since expanded.

International Terrestrial Reference System (ITRS)
   A very precise, geocentric coordinate system. The ITRS is geocentric, including the oceans and the atmosphere. Its length unit is the meter. Its axes are consistent with the Bureau International de l'Heuer (BIH) System at 1984.0 within ± 3 milliarcseconds. The IERS Reference Pole (IRP) and Reference Meridian (IRM) are consistent with the corresponding directions in the BIH directions within ± 0.005". The BIH reference pole was adjusted to the Conventional International Origin (CIO) in 1967 and kept stable independently until 1987. The uncertainty between IRP and CIO is ± 0.03". The ITRS is realized by estimates of the coordinates and velocities of a set of stations some of which are Satellite Laser Ranging (SLR) stations, or Very Long Baseline Interferometry (VLBI) stations, but the vast majority are GPS tracking stations of the IGS network.

Ionosphere
   A layer of atmosphere extending from about 50 to 1000 kilometers above the Earth's surface in which gas molecules are ionized by ultraviolet radiation from the sun. The apparent speed, polarization, and direction of GPS signals are affected by the density of free electrons in this nonhomogeneous and dispersive band of atmosphere.

Ionospheric Delay
(aka Ionospheric Refraction)
   The difference in the propagation time for a signal passing through the ionosphere compared with the propagation time for the same signal passing through a vacuum. The magnitude of the ionospheric delay changes with the time of day, the latitude, the season, solar activity, and the observing direction. For example, it

is usually least at the zenith and increases as a satellite gets closer to the horizon. The ionospheric delay is frequency dependent and, unlike the troposphere, affects the L1 and L2 carriers differently. There are two categories of the ionospheric delay, phase and group. Group delay affects the codes, the modulations on the carriers; phase advance affects the carriers themselves. Group delay and phase advance have the same magnitude but opposite signs. Since the ionospheric delay is frequency dependent it can be nearly eliminated by combination of pseudorange or carrier phase observations on both the L1 and L2 carriers. Still, even with dual frequency observation and relative positioning methods in place, over long baselines the residual ionospheric delay can remain a substantial bias for high precision GPS. Single frequency receivers cannot significantly mitigate the error at all and must depend on the ionospheric correction available in the Navigation Message to remove even 50% of the effect.

Isotropic Antenna
An isotropic antenna is a hypothetical, lossless antenna that has equal capabilities in all directions.

Iteration
Converging on a solution by repetitive operations.

IVHS
Intelligent Vehicle Highway System.

Joint Program Office (JPO)
The office responsible for the management of the GPS system in the U.S. Department of Defense.

Kalman Filter
In GPS a numerical data combiner used in determining an instantaneous position estimate from multiple statistical measurements on a time-varying signal in the presence of noise. The Kalman filter is a set of mathematical equations that provides an estimation technique based on least squares. This recursive solution to the discrete-data linear filtering problem was proposed by R.E. Kalman in 1960. Since then the Kalman filter has been applied to radionavigation in general and GPS in particular, among other methods of measurement.

Kinematic Positioning
(aka Stop & Go Positioning)
(also see Real-Time Kinematic Positioning)
(1) A version of relative positioning in which one receiver is a stationary reference and at least one other roving receiver coordinates unknown positions with short occupation times while both track the same satellites and maintain constant lock. If lock is lost reinitialization is necessary to fix the integer ambiguity.
(2) GPS applications in which receivers on vehicles are in continuous motion.

L-Band
Radio frequencies from 390 MHz to 1550 MHz.

L1
A GPS signal with the C/A code, the P code, and the Navigation Message modulated onto a carrier with the frequency 1575.42 MHz.

L2
A GPS signal with the P code, and the Navigation Message modulated onto a carrier with the frequency 1227.60 MHz.

Latency
The time taken for a system to compute corrections and transmit them to users in real-time GPS.

Latitude
An angular coordinate, the angle measured from an equatorial plane to a line. On Earth the angle measured northward from the equator is positive, southward is negative. The geodetic latitude of a point is the angle between the equatorial plane of the ellipsoid and a normal to the ellipsoid through the point. At that point the astronomic latitude differs from the geodetic latitude by the meridional component of the deflection of the vertical.

Longitude
An angular coordinate. On Earth, the dihedral angle from the plane of reference, 0° meridian, to a plane through a point of concern, and both planes perpendicular to the plane of the equator.

In 1884, the Greenwich Meridian was designated the initial meridian for longitudes. The geodetic longitude of a point is the angle between the plane of the geodetic meridian through the point and the plane of the 0° meridian. At that point the astronomic longitude differs from the geodetic longitude by the amount of the component in the prime vertical of the local deflection of the vertical divided by the cosine of the latitude.

Loop Closure
A procedure by which the internal consistency of a GPS network is discovered. A series of baseline vector components from more than one GPS session, forming a loop or closed figure, is added together. The closure error is the ratio of the length of the line representing the combined errors of all the vectors' components to the length of the perimeter of the figure.

Mask Angle
An elevation angle below which satellites are not tracked. The technique is used to mitigate atmospheric, multipath, and attenuation errors. A usual mask angle is 15°.

Master Control Station
A facility manned by the 2nd Space Operations Squadron at Schriever Air Force Base in Colorado Springs, Colorado. The Master Control Station is the central facility in a network of worldwide tracking and upload stations that comprise the GPS Control Segment.

MCS
GPS Master Control Station.

Megahertz (MHz)
One million cycles per second.

Microsecond (μsec)
One millionth (10^6) of a second.

Millisecond (ms or msec)
One thousandth of a second.

Minimally Constrained
(see Network Adjustment)
A least squares adjustment of all observations in a network with only the constraints necessary to achieve a meaningful solution. For example, the adjustment of a GPS network with the coordinates of only one station fixed.

Modem
(Modulator/Demodulator)
A device that converts digital signals to analog signals and analog signals to digital signals. Computers sharing data usually require a modem at each end of a phone line to perform the conversion.

Monitor Stations
Stations used in the GPS control segment to track satellite clock and orbital parameters. Data collected at monitor stations are linked to a master control station at which corrections are calculated and from which correction data are uploaded to the satellites as needed.

Multichannel Receiver
(aka Parallel Receiver)
A receiver with many independent channels. Each channel can be dedicated to tracking one satellite continuously.

Multipath
(aka Multipath Error)
The error that results when a portion of the GPS signal is reflected. When the signal reaches the receiver by two or more different paths, the reflected paths are

longer and cause incorrect pseudoranges or carrier phase measurements and subsequent positioning errors. Multipath is mitigated with various preventative antenna designs and filtering algorithms.

Multiplexing Channel
(aka Fast-switching, Fast-sequencing, Fast-multiplexing)

A channel of a GPS receiver that tracks through a series of satellites' signals, from one signal to the next in a rapid sequence.

Multiplexing Receiver

A GPS receiver that tracks satellites' signals sequentially and differs from a Multichannel Receiver in which individual channels are dedicated to each satellite signal.

NAD 83
(North American Datum, 1983)

The horizontal control datum for positioning in Canada, the United States, Mexico, and Central America based on a geocentric origin and the Geodetic Reference System 1980 (GRS80) ellipsoid. The values for GRS 80 adopted by the International Union of Geodesy and Geophysics in 1979 are, a = 6378137 meters and reciprocal of flattening = 1/f = 298.257222101.

It was designed to be compatible with the Bureau International de l'Heuer (BIH) Terrestrial System BTS84. The ellipsoid is geocentric. The origin was defined by satellite laser ranging (SLR), orientation by astronomic observations and scale by both SLR and very long baseline interferometry (VLBI).

NAD83 was actually realized through Doppler observations using internationally accepted transformations from the Doppler reference frame to BTS84 and adjustment of some 250,000 points. VLBI stations were also included to provide an accurate connection to other reference frames.

While NAD 83 is similar to the World Geodetic System of 1984 (WGS 84), it is not the same as WGS 84. Defined and maintained by the U.S. Department of Defense, WGS84 is a global geodetic datum used by the GPS Control Segment.

Access to NAD83 is through a national network of horizontal control monuments established mainly by conventional horizontal control methods, triangulation, trilateration, and astronomic azimuths. Some GPS baselines were used. The adjustment of these horizontal observations, together with several hundred observed Doppler positions, provides a practical realization of NAD83.

Nanosecond (ns or nsec)

One billionth (10^9) of a second.

NANU

Notice Advisory to NAVSTAR Users.

North American Vertical Datum of 1988 (NAVD 88)

A minimally constrained adjustment of U.S., Canadian, and Mexican leveling observations holding fixed the height of the primary tidal benchmark of the new International Great Lakes Datum of 1985 (IGLD 85) at Father Point/Rimouski, Quebec, Canada. NAVD88 and IGLD 85 are now the same. Between NAVD88 orthometric heights and those referred to the National Geodetic Vertical Datum of 1929 (NGVD 29) there are differences ranging from –40 cm to + 150 cm within the lower 48 states. The differences range from + 94 cm to + 240 cm in Alaska. GPS derived orthometric heights estimated using the precise geoid models now available are compatible with NAVD88. NAVD88 includes 81,500 km of new leveling data never before adjusted to NGVD 29. The principal impetus for NAVD88 was minimizing the recompilation of national mapping products. The NAVD 88 datum does not correspond exactly to the theoretical level surface defined by the GRS80 definitions.

Navigation Message

(aka Data Message)

A message modulated on L1 and L2 of the GPS signal that includes an ionospheric model, the satellite's broadcast ephemeris, broadcast clock correction, constellation almanac and health among other things.

NAVSAT

A European radionavigation system under development.

NAVSTAR

(NAVigation Satellite Timing and Ranging)

The GPS satellite system.

Network Adjustment

A least-squares solution in which baselines vectors are treated as observations. It may be minimally constrained. It may be constrained by more than one known coordinate, as is usual in a GPS survey to densify previously established control or a geodetic framework.

Observing Session

(aka Observation)

Continuous and simultaneous collection of GPS data by two or more receivers.

Omega

A radionavigation system that can provide global coverage with only eight ground-based transmitting stations.

On-the-Fly (OTF)

A method of resolving the carrier phase ambiguity very quickly. The method requires dual-frequency GPS receivers capable of making both carrier phase and precise pseudorange measurements. The receiver is not required to remain stationary, making the technique useful for initializing in carrier phase kinematic GPS.

Orthometric Height

The vertical distance from the geoid to the surface of the Earth. GPS heights are ellipsoidal. Ellipsoidal heights are the vertical distance from an ellipsoid of reference to the Earth's surface. Ellipsoidal heights differ from leveled, orthometric heights. And conversion from ellipsoidal to orthometric heights requires the vertical distance from the ellipsoid of reference to the geoid, the geoidal height. The vertical distance from the ellipsoid of reference to the geoid around the world varies from + 75 to –100 meters; in the conterminous United States it varies from –8 meters to –53 meters. The geoidal heights are negative because the geoid is beneath the ellipsoid. In other words, the ellipsoid is overhead.

The relationship between these three heights is:

$$h = H + N$$

Where h is the ellipsoid height, H is the orthometric height, and N is the geoidal height.

Outage

GPS positioning service is unavailable. Possible reasons for an outage include sufficient number of satellites are not visible, the dilution of precision value is too large, or the signal to noise ratio value is too small.

Parallel Receiver

(see Multichannel Receiver)

P Code

(aka Protected Code)

A binary code known by the names P-Code, Precise Code, and Protected Code. It is a standard spread-spectrum GPS pseudorandom noise code. It is modulated on the L1 and the L2 carrier using binary biphase modulations. Each week-long segment of the P Code is unique to a single GPS satellite and is repeated each week. It has a chipping rate 10.23 MHz, more than ten million bits per second. The P Code is sometimes replaced with the more secure Y Code in a process known as Antispoofing.

PDOP–Position Dilution of Precision

(see Dilution of Precision)

Perturbation

Deviation in the path of an object in orbit, deviation being departure of the actual orbit from the predicted Keplerian orbit. Perturbing forces of earth-orbiting satellites are caused by atmospheric drag, radiation pressure from the sun, the gravity of the moon and the sun, the geomagnetic field, and the noncentral aspect of earth's gravity.

Phase Lock

The adjustment of the phase of an oscillator signal to match the phase of a reference signal. First, the receiver compares the phases of the two signals. Next, using the resulting phase difference signal, the reference oscillator frequency is adjusted. When the two signals are next compared the phase difference between them is eliminated.

Phase Observable

(see Carrier Phase)

Phase Smoothed Pseudorange

A pseudorange measurement with its random errors reduced by combination with carrier phase information.

Picosecond

One trillionth (10^{12}) of a second, one millionth of a microsecond.

Point Positioning

(see Absolute Positioning)

Position

A description, frequently by coordinates, of the location and orientation of a point or object.

Postprocessed GPS

A method of deriving positions from GPS observations in which base and roving receivers do not communicate as they do in Real-Time Kinematic (RTK) GPS. Each receiver records the satellite observations independently. Their collections are combined later. The method can be applied to pseudoranges to be differentially corrected or carrier phase measurements to be processed by double-differencing.

Precise Code

(see P-Code)

Precise Ephemeris

(see Ephemeris)

Precise Positioning Service (PPS)

GPS positioning for the military at a higher level of absolute positioning accuracy than is available to C/A code receivers, which relies on SPS, Standard Positioning Service. PPS is based on the dual frequency P code.

Precision

Agreement among measurements of the same quantity, widely scattered results are less precise than those that are closely grouped. The higher the precision, the smaller the random errors in a series of measurements. The precision of a GPS survey depends on: the network design, surveying methods, processing procedures, and equipment.

Protected Code

(see P Code)

Pseudolite

(aka Pseudo Satellite)

A ground-based differential station which simulates the signal of a GPS satellite with a typical maximum range of 50 km. Pseudolites can enhance the accuracy and extend the coverage of the GPS constellation. Pseudolite signals are designed to minimize their interference with the GPS signal.

Pseudorandom Noise

(aka PRN)

A sequence of digital 1's and 0's that appear to be randomly distributed, but can be reproduced exactly. Binary signals with noise-like properties are modulated on the GPS carrier waves as the C/A codes and the P codes. Each GPS satellite has unique C/A and P codes. A satellite may be identified according to its PRN number. Thirty-two GPS satellite pseudorandom noise codes are currently defined.

Pseudorange

In GPS, a time-biased distance measurement. It is based on code transmitted by a GPS satellite, collected by a GPS receiver, and then correlated with a replica of the same code generated in the receiver. However, there is no account for errors in synchronization between the satellite's clock and the receiver's clock in a pseudorange. The precision of the measurement is a function of the resolution of the code, therefore a C/A code pseudorange is less precise than a P code pseudorange.

Quartz Crystal Controlled Oscillator

GPS receivers rely on a quartz crystal oscillator to provide a stable reference so that other frequencies of the system can be compared with or generated from this reference. The fundamental component is the quartz crystal resonator. It utilizes the piezoelectric effect. When an electrical signal is applied the quartz resonates at a frequency unique to its shape, size, and cut. The first study of the use of quartz crystal resonators to control the frequency of vacuum tube oscillators was made by Walter G. Cady in 1921. Important contributions were made by G. W. Pierce, who showed that plates of quartz cut in a certain way could be made to vibrate so as to control frequencies proportional to their thickness.

R95
A representation of positional accuracy. The radius of a theoretical circle centered at the true position that would enclose 95 % of the other positions.

Radionavigation
The determination of position, direction, and distance using the properties of transmitted radio waves.

Range
(aka Geometric Range)

A distance between two points, particularly the distance between a GPS receiver and satellite.

Range Rate
The rate at which the range between a GPS receiver and satellite changes. Usually measured by tracking the variation in the Doppler Shift.

Real-Time DGPS
A method of determining relative positions between known control and unknown positions using pseudorange measurements. A base station or satellite at the known position transmits corrections to the roving receiver or receivers. The procedure offers less accuracy than RTK. However, the results are immediately available, in real-time, and need not be postprocessed.

Real-Time Kinematic (RTK)
A method of determining relative positions between known control and unknown positions using carrier phase measurements. A base station at the known position transmits corrections to the roving receiver or receivers. The procedure offers high accuracy immediately, in real-time. The results need not be postprocessed. In the earliest use of GPS, kinematic and rapid static positioning were not frequently used because ambiguity resolution methods were still inefficient. Later when ambiguity resolution such as on-the-fly (OTF) became available, real-time kinematic and similar surveying methods became more widely used.

Receiver Channel
(see Channel)

Reconstructed Carrier Phase
(see Carrier Phase)

Reference Station
(see Base Station)

Relative Positioning
A GPS surveying method that improves the precision of carrier phase measurements. One or more GPS receivers occupy a base station, a known position. They

collect the signals from the same satellites at the same time as other GPS receivers that may be stationary or moving. The other receivers occupy unknown positions in the same geographic area. Occupying known positions, the base station receivers find corrective factors that can either be communicated in real-time to the other receivers, as in RTK, or may be applied in postprocessing, as in static positioning. Relative positioning is in contrast to Absolute, or Point, Positioning. In relative positioning, errors that are common to both receivers, such as satellite clock biases, ephemeris errors, propagation delays, etc., are mitigated.

RINEX
(Receiver Independent Exchange Format)

A package of GPS data formats and definitions that allow interchangeable use of data from dissimilar receiver models and postprocessing software developed by the Astronomical Institute of the University of Berne in 1989. More than 60 receivers from four different manufacturers were used in the GPS survey EUREF 89. RINEX was developed for the exchange of the GPS data collected in that project.

Root Mean Square (RMS)

The square root of the mean of squared errors for a sample.

Rover
(aka Mobile Receiver)

A GPS receiver that moves relative to a stationary base station during a session.

RTCM
(Radio Technical Commission on Maritime services)

In DGPS the abbreviation RTCM has come to mean the correction messages transmitted by some reference stations using a protocol developed by the Radio Technical Commission on Maritime services Special Committee 104. These corrections can be collected and decoded by DGPS receivers designed to accept the signal. The corrections allow the receiver to generate corrected coordinates in real-time. There are several sources of RTCM broadcasts. One source is the U.S. Coast Guard system of beacons in coastal areas. Other sources are commercial services, some of which broadcast RTCM corrections by satellite, some use FM subcarriers.

Rubidium Clock

An environmentally tolerant and very accurate atomic clock whose working element is gaseous rubidium. The resonant transition frequency of the Rb-87 atom (6,834,682,614 Hz) is used as a reference. Rubidium frequency standards are small, light, and have low power consumption.

SA
(see Selective Availability)

S-Code
(see C/A Code)

Satellite Clocks
Two rubidium (Rb) and two cesium (Cs) atomic clocks are aboard Block II/IIA satellites. Three rubidium clocks are on Block IIR satellites. Hydrogen Maser time standards may be used in future satellites.

Satellite Constellation
In GPS, four satellites in each of six orbital planes. In GLONASS, eight satellites in each of three orbital planes.

Second
Base unit of time in the International System of Units. The duration of 9,192,631,770 periods of the radiation corresponding to the transition between the two hyperfine levels of the ground state of the cesium-133 atom undisturbed by external fields.

Selective Availability (SA)
By dithering the timing and ephemrides data in the satellites, standard positioning service (SPS) users are denied access to the full GPS accuracy. The intentional degradation of the signals transmitted from the satellites may be removed by precise positioning service (PPS) users. It is also virtually eliminated in carrier phase relative positioning techniques and double-differencing processing. Selective availability was begun in 1990 and was removed from August 10, 1990 to July 1, 1991. It was reenabled on November 15, 1991 and finally switched off by presidential order on May 2, 2000.

Single Difference
(see Between-Receiver Single Difference)
(also see Between-Satellite Single Difference)

Space Segment
The portion of the GPS system in space, including the satellites and their signals.

Spatial Data
(aka Geospatial Data)
Information that identifies the geographic location and characteristics of natural or constructed features and boundaries of earth. This information may be derived from, among other things, remote sensing, mapping, and surveying technologies.

Special Theory of Relativity

(see also General Theory of Relativity)

A theory developed by Albert Einstein, predicting, among other things, the changes that occur in length, mass, and time at speeds approaching the speed of light.

General relativity predicts that as gravity weakens the rate of clocks increase, they tick faster. On the other hand, special relativity predicts that moving clocks appear to tick more slowly than stationary clocks, since the rate of a moving clock seems to decrease as its velocity increases. Therefore, for GPS satellites, general relativity predicts that the atomic clocks in orbit on GPS satellites tick faster than the atomic clocks on earth by about 45,900 ns/day. Special relativity predicts that the velocity of atomic clocks moving at GPS orbital speeds causes them to tick slower by about 7,200 ns/day than clocks on earth. The rates of the clocks in GPS satellites are reset before launch to compensate for these predicted effects.

Spherical Error Probable (SEP)

A description of three-dimensional precision. The radius of a sphere with its center at the actual position, that is expected to be large enough to include half (50 %) of the normal distribution of the scatter of points observed for the position.

Spread-Spectrum Signal

A signal spread over a frequency band wider than needed to carry its information. In GPS a spread-spectrum signal is used to prevent jamming, mitigate multipath, and allow unambiguous satellite tracking.

SPS

(see Standard Positioning Service)

Standard Deviation

(aka 1 sigma, 1σ)

An indication of the dispersion of random errors in a series of measurements of the same quantity. The more tightly grouped the measurements around their average (mean), the smaller the standard deviation. Approximately 68 % of the individual measurements will be within the range expressed by the standard deviation.

Standard Positioning Service (SPS)

Civilian absolute positioning accuracy using pseudorange measurements from a single frequency C/A code receiver, ± 20 to ± 40 meters 95 % of the time (2 drms) with Selective Availability switched off.

Static Positioning

A relative, differential, surveying method in which at least two stationary GPS receivers collect signals simultaneously from the same constellation of satellites during long observation sessions. Generally, static GPS measurements are post-processed and the relative position of the two units can be determined to a very high accuracy. Static positioning is in contrast to kinematic and real-time kinematic positioning where one or more receivers track movement. Static positioning is in contrast to absolute, or point, positioning, which has no relative positioning component. Static positioning is in contrast to rapid-static positioning where the observation sessions are short.

Stop-and-Go Positioning
(see Kinematic Positioning)

SV
Space Vehicle.

Time Dilation
(aka Relativistic Time Dilation)
(see Special Theory of Relativity)
Systematic variation in time's rate on an orbiting GPS satellite relative to time's rate on earth. The variation is predicted by the special theory of relativity and the general theory of relativity as presented by A. Einstein.

Transit Navigation System
A satellite based Doppler positioning system funded by the U.S. Navy.

Triple-Difference
(aka Receiver-Satellite-Time Triple difference)
The combination of two double differences. Each of the double differences involves two satellites and two receivers. The triple difference is derived between two epochs. In other words, a triple difference involves two satellites, two receivers, and time. Triple differences ease the detection of cycle slips.

Trivial Baseline
Baselines observed using GPS Relative Positioning techniques. When more than two receivers are observing at the same time, both independent (nontrivial) and trivial baselines are generated. For example, where r is the number of receivers, every complete static session yields r-1 independent (nontrivial) baselines and the remaining baselines are trivial. If four receivers are used simultaneously, six baselines are created. Three are trivial and three are independent. The three independent baselines will fully define the position of each occupied station in relation to the others. The three trivial baselines may be processed, but the observational data used for them has already produced the independent baselines. Therefore, only the independent baseline results should be used for the network adjustment or quality control.

Tropospheric Effect
(aka Tropospheric Delay)
The troposphere comprises approximately 9 km of the atmosphere over the poles and 16 km over the equator. The tropospheric effect is nondispersive for

frequencies under 30 GHz. Therefore, it affects both L1 and L2 equally. Refraction in the troposphere has a dry component and a wet component. The dry component is related to the atmospheric pressure and contributes about 90% of the effect. It is more easily modeled than the wet component. The GPS signal that travels the shortest path through the atmosphere will be the least affected by it. Therefore, the tropospheric delay is least at the zenith and most near the horizon. GPS receivers at the ends of short baselines collect signals that pass through substantially the same atmosphere and the tropospheric delay may not be troublesome. However, the atmosphere may be very different at the ends of long baselines.

UKOOA

Member companies of the UK Offshore Operators Association (UKOOA) are licensed by the government to search for and produce oil and gas in UK waters. It publishes, among other things "The Use of Differential GPS in Offshore Surveying." These guidelines cover installation and operation, quality measures, minimum training standards, receiver outputs, and data exchange format.

User-Equivalent Range Error (UERE)

The contribution in range units of individual uncorrelated biases to the range measurement error.

USCG

United States Coast Guard.

Universal Transverse Mercator (UTM)

The transverse Mercator projection represents ellipsoidal positions, latitude and longitude, as grid coordinates, northing and easting, on a cylindrical surface that can be developed into a flat surface. Universal Transverse Mercator is a particular type of transverse Mercator projection. It was adopted by the U.S. Army for large-scale military maps and is shown on all 7.5 minute quadrangle maps and 15 minute quadrangle maps prepared by the U.S. Geological Survey. The earth is divided into 60 zones between 84°N. latitude and 84°S. latitude, most of which are 6° of longitude wide. Each of these UTM zones have a unique central meridian and the scale varies by 1 part in 1,000 from true scale at equator.

User Interface

The software and hardware that activate displays and controls that are the means of communication between a GPS receiver and the receiver's operator.

User Segment

That component of the GPS system that includes the user equipment, applications, and operational procedures.

The part of the whole GPS system that includes the receivers of GPS signals.

UTC
(see Coordinated Universal Time)

UTM
(see Universal Transverse Mercator)

Very Long Baseline Interferometry (VLBI)
By measuring the arrival time of the wavefront emitted by a distant quasar at two widely separated earth-based antennas, the relative position of the antennas is determined. Because the time difference measurements are precise to a few picoseconds, the relative positions of the antennas are accurate within a few millimeters and the quasar positions to fractions of a milliarcsecond.

Voltage-Controlled Quartz Crystal Oscillator
(see also Quarts Crystal Controlled Oscillator)
A quartz crystal oscillator with a voltage controlled frequency.

Wavelength
(see also Frequency)
Along a sine wave the distance between adjacent points of equal phase. The distance required for one complete cycle.

Waypoint
A two-dimensional coordinate to be reached by GPS navigation.

Wide Area Augmentation System (WAAS)
A U.S. Federal Aviation Authority (FAA) system that augments GPS accuracy, availability, and integrity. It provides a satellite signal for users to support en route and precision approach aircraft navigation. Similar systems are Europe's European Geostationary Navigation Overlay System and Japan's MT-SAT.

World Geodetic System 1984 (WGS84)
A world geodetic earth-centered, earth-fixed terrestrial reference system. The origin of the WGS 84 reference frame is the center of mass of the earth. The GPS Control Segment has worked in WGS84 since January of 1987 and therefore, GPS positions are said to be in this datum.

XTE
(see Crosstrack Error)

Y-Code
When Antispoofing is on, the P code is encrypted into the Y code and transmitted on L1 and L2.

Zero Baseline Test

A setup using two GPS receivers connected to one antenna. Nearly all biases are identical for both receivers and only random observation errors attributable to both receivers remain. The baseline measured should be zero if the receivers' calibration were ideal.

Z-Count Word

The GPS satellite clock time at the leading edge of the next data subframe of the transmitted GPS message (usually expressed as an integer number of 1.5 second periods) [van Dierendock et al., 1978].

Zulu Time

(see Coordinated Universal Time)

References

Brunner, F.K., and W.M. Welsch. 1993. "Effect of the Troposphere on GPS Measurements," in *GPS World,* Vol. 4, No. 1, Advanstar Communications, pp. 42–51.

Federal Geodetic Control Committee. 1989. *Geometric Geodetic Accuracy Standards and Specifications For Using GPS Relative Positioning Techniques,* Version 5, Aug. 1, 1989.

Kalman, R.E. 1960. "A New Approach to Linear Filtering and Prediction Problems," Trans. of the ASME, *Journal of Basic Engineering,* pp. 35–45.

Kaplan, E.D. 1996. "Understanding GPS, Principles and Applications," Artech House Publishers, Norwood, MA, pp. 186–189.

Langley, R.B. 1991. "The GPS Receiver: An Introduction," in *GPS World,* Vol. 2, No. 1, Advanstar Communications, pp. 50–53.

Langley, R.B. 1993. "The GPS Observables," in *GPS World,* Vol. 4, No. 4, Advanstar Communications, pp. 52–59.

Langley, R.B. (1998). "RTK GPS," *Innovation* in *GPSWorld,* September 1998.

Leick, A. "The Least-Squares Toolbox," *ACSM Bulletin,* December 1993.

National Geodetic Survey. 1986. "Geodetic Glossary," NOAA Technical Publications, U.S. Department of Commerce, p. 209.

Reilly, J.P. 1996. "GPS Calibration," *The GPS Observer* in *Point of Beginning,* April 1996, pp. 20–23.

Reilly, J.P. 1997. "Elevations from GPS," *The GPS Observer* in *Point of Beginning,* May 1997, pp. 24–26.

Reilly, J.P. 2000. "GEOID99," *The GPS Observer* in *Point of Beginning,* January 2000, pp. 12–13.

Trimble Navigation Limited. *WAVE Baseline Processor and SV Azimuth & Elevation Table.* GPSurvey Software Suite. Version 1.10D.

Trimble Navigation Limited. *Model 4000ST GPS Surveyor, Operation Manual,* July 1989 3.2X, p. D-1.

U.S. Department of Commerce, NOAA, National Ocean Survey, National Geodetic Survey. *Control Data Sheets.*

U.S. Department of Commerce, NOAA, National Ocean Survey, National Geodetic Survey. *DSDATA format specifications,* www.ngs.noaa.gov.

van Dierendock, A.J., S.S. Russell, E.R. Kopitzke, and M. Birnbaum. 1978. "The GPS Navigation Message," in *Global Positioning System.* Papers published in *Navigation,* reprinted by *The Institute of Navigation,* Vol. I, 1980, pp. 55–73.

Wells, D., Ed. 1986. *Guide to GPS Positioning,* Canadian GPS Associates, Fredericton, New Brunswick.

Index